SpringerBriefs in Applied Sciences and Technology

SpringerBriefs present concise summaries of cutting-edge research and practical applications across a wide spectrum of fields. Featuring compact volumes of 50 to 125 pages, the series covers a range of content from professional to academic.

Typical publications can be:

- A timely report of state-of-the art methods
- An introduction to or a manual for the application of mathematical or computer techniques
- A bridge between new research results, as published in journal articles
- A snapshot of a hot or emerging topic
- An in-depth case study
- A presentation of core concepts that students must understand in order to make independent contributions

SpringerBriefs are characterized by fast, global electronic dissemination, standard publishing contracts, standardized manuscript preparation and formatting guidelines, and expedited production schedules.

On the one hand, **SpringerBriefs in Applied Sciences and Technology** are devoted to the publication of fundamentals and applications within the different classical engineering disciplines as well as in interdisciplinary fields that recently emerged between these areas. On the other hand, as the boundary separating fundamental research and applied technology is more and more dissolving, this series is particularly open to trans-disciplinary topics between fundamental science and engineering.

Indexed by EI-Compendex, SCOPUS and Springerlink.

Manish Kumar Goyal · Kuldeep Singh Rautela

Aerosol Atmospheric Rivers

Availability, Spatiotemporal Characterisation, Predictability, and Impacts

 Springer

Manish Kumar Goyal 🆔
Department of Civil Engineering
Indian Institute of Technology Indore
Indore, Madhya Pradesh, India

Kuldeep Singh Rautela 🆔
Department of Civil Engineering
Indian Institute of Technology Indore
Indore, Madhya Pradesh, India

ISSN 2191-530X   ISSN 2191-5318   (electronic)
SpringerBriefs in Applied Sciences and Technology
ISBN 978-3-031-66757-2   ISBN 978-3-031-66758-9   (eBook)
https://doi.org/10.1007/978-3-031-66758-9

This Springer imprint is published by the registered company Springer Nature Switzerland AG
The registered company address is: Gewerbestrasse 11, 6330 Cham, Switzerland

If disposing of this product, please recycle the paper.

Preface

Aerosol pollution, with its diverse sources and significant impacts on climate, weather, and health, poses a complex environmental challenge. Among the various phenomena contributing to this issue, Aerosol Atmospheric Rivers (AARs) have emerged as a critical area of study. AARs, similar to atmospheric rivers (ARs) that transport moisture, carry concentrated aerosols across vast distances, affecting regions far beyond their origin. Understanding the dynamics and impacts of AARs is essential for improving our predictive capabilities and developing effective mitigation strategies.

In recent years, integrating data mining, artificial intelligence (AI), and machine learning (ML) into environmental science has revolutionized our ability to study and assess aerosol pollution. These advanced computational techniques enable us to process large datasets, uncover hidden patterns, and create robust predictive models. This book explores this cutting-edge intersection, showcasing the potential of AI and ML in enhancing our understanding towards the transportation of aerosols.

The book provides a comprehensive overview of aerosol science, starting with the fundamental concepts and moving into the specifics of AARs, including their detection, monitoring, and impacts on air quality and health. Through detailed discussions and case studies, it highlights the transformative role of AI and ML in forecasting aerosol concentrations, identifying pollution hotspots, and predicting the spatio-temporal patterns of AARs.

This book aims to equip researchers, policymakers, and practitioners with the knowledge and tools to tackle aerosol pollution. Harnessing the power of artificial intelligence and fostering interdisciplinary collaboration, accurate predictions, effective mitigation strategies, and ultimately safeguarding public health and environmental quality may be developed.

Indore, India
Manish Kumar Goyal
Kuldeep Singh Rautela

Contents

Chapter 1
Understanding the Significance of Aerosol Pollution and Aerosol Atmospheric Rivers

Abstract The air quality, weather patterns, and temperature of Earth are all significantly impacted by aerosols, which are tiny particles suspended in the atmosphere. Aerosols come in a wide range of sizes and compositions from both natural and man-made sources, ranging from mineral dust to pollutants released during industrial processes. Aerosols have a significant impact on ecosystems, local air quality, and climate trends due to their complex transport mechanisms, which include turbulent mixing and atmospheric circulation patterns. Aerosol pollution also affects clouds in a variety of ways, causing changes in their characteristics, absorbing, and reflecting solar energy, and aggravating respiratory and cardiovascular conditions. Mitigation solutions for this ubiquitous environmental concern include technology advancements, governmental initiatives, and behavioural changes. Aerosol emissions have been notably reduced by laws like the Clean Air Act in the United States, and particulate matter from industrial pollutants can be effectively removed by devices like electrostatic precipitators. Clean cooking techniques are also promoted by public awareness efforts, which help to lower aerosol pollution both indoors and outdoors. Additionally, the recently developed idea of aerosol atmospheric rivers (AARs) reveals lengthy, narrow passageways of concentrated aerosol particles that are driven by complex atmospheric dynamics and synoptic-scale circulation patterns through the atmosphere. AARs are crucial for advancing predictive capabilities and implementing effective mitigation strategies to combat aerosol pollution and its far-reaching impacts on climate, weather, and public health.

Keywords Aerosol pollution · Air quality · Aerosol atmospheric rivers (AARs) · Public health

1.1 Introduction

Aerosols, tiny particles suspended in Earth's atmosphere, play a pivotal role in shaping our planet's climate, weather patterns, and air quality (Kuniyal and Guleria 2019). Aerosols are extremely diverse in terms of their sources and composition,

© The Author(s), under exclusive license to Springer Nature Switzerland AG 2024
M. K. Goyal and K. S. Rautela, *Aerosol Atmospheric Rivers*,
SpringerBriefs in Applied Sciences and Technology,
https://doi.org/10.1007/978-3-031-66758-9_1

and they range in size from a few nanometres to several micrometres (Pryor et al. 2015). Aerosols encompass various particle types, each with distinct characteristics and origins (Seinfeld 2015). Primary aerosols, which come from natural sources such as mineral dust particles carried by wind from arid regions and sea salt aerosols produced by ocean spray, are released directly into the atmosphere (Tomasi and Lupi 2017). Air pollutants like sulphur dioxide, nitrogen oxides, and particulate matter are released into the atmosphere by human activities like industrial processes, burning fossil fuels, and agricultural practices (Amann et al. 2020). These activities are known as anthropogenic primary aerosols. Ammonia, nitrogen oxides, sulphur dioxide, volatile organic compounds, and other precursor gases react chemically in the atmosphere to produce secondary aerosols (Chen et al. 2019). These processes result in the production of organic, secondary organic, nitrate, and sulphate aerosols, which add to a load of anthropogenic and natural aerosols. Aerosols travel through a variety of intricate processes once they are released, such as turbulent mixing, atmospheric convection, and large-scale circulation patterns, all of which are impacted by atmospheric dynamics. Long-term suspension in the atmosphere allows fine aerosol particles to travel great distances over continents and even oceans (Adebiyi et al. 2023). Jet streams and trade winds are examples of atmospheric circulation systems that are important for the global redistribution of aerosols, which impacts ecosystem dynamics, regional air quality, and climate trends (Goyal et al. 2012; Chakraborty et al. 2016). Weather events like dust storms, volcanic eruptions, and wildfires can also increase aerosol transport because they release a lot of aerosol particles into the atmosphere, which promotes long-distance movement and extensive dispersion (Wang and Strong 2019; Sarkar et al. 2019). Accurately evaluating aerosol transportation processes is necessary to determine their effects on local and global climate, weather, and air quality.

Aerosol pollution has a wide range of diverse effects (Goyal and Ojha 2011; Koplitz et al. 2016; Sarkar et al. 2019). Aerosol particles scatter and absorb solar radiation, causing changes to the Earth's energy balance and local temperature trends (Pöschl 2005). These are known as direct impacts. Aerosol-cloud interactions have indirect impacts that affect precipitation processes, cloud formation, and characteristics (Silva et al. 2020). Aerosols can also deteriorate the quality of the air, which increases the risk of cardiovascular disease, respiratory ailments, and early death (Southerland et al. 2022). Aerosols and atmospheric processes interact in a complicated way, which emphasizes how crucial it is to comprehend their origins, compositions, and effects on weather, climate, and public health (Hidy 2019; Jha et al. 2019; Sharma and Goyal 2020; Kuniyal et al. 2021; Kumar et al. 2021b, a; Poonia et al. 2021).

Furthermore, atmospheric scientists are becoming increasingly interested in the idea of aerosol atmospheric rivers (AARs) (Chakraborty et al. 2021, 2022; Rautela et al. 2024a). Aerosol atmospheric rivers are extended passageways of concentrated aerosol particles that travel great distances across the atmosphere (Chakraborty et al. 2022). While sharing characteristics with moisture-based atmospheric rivers, aerosol atmospheric rivers are distinct in their composition and impacts on atmospheric dynamics and weather patterns (Singh et al. 2023; Singh and Goyal 2023a, b). Understanding the formation, dynamics, and implications of aerosol atmospheric rivers is crucial for advancing our knowledge of atmospheric processes and their interactions with aerosol particles.

Aerosol pollution mitigation calls for coordinated efforts on several fronts, including governmental interventions, technology developments, and behavioural adjustments (Burns et al. 2019). Anthropogenic aerosol emissions from the transportation, energy, and industrial sectors can be decreased with the support of regulatory measures including pollution controls, emissions standards, and incentives for the adoption of clean energy (Aminzadegan et al. 2022). Investments in sustainable practices and cleaner technology research and development are crucial for reducing aerosol pollution while fostering economic expansion and environmental sustainability.

1.2 Aerosol Pollution: Sources and Composition

Aerosol pollution is a result of both natural and human activities, and it poses a major threat to the environment because of its wide-ranging effects on the climate, air quality, and human health (Li et al. 2017). Understanding the origins and makeup of aerosols is imperative in evaluating their impacts and executing efficacious measures to mitigate them (Fig. 1.1). An extensive examination of aerosol pollution is given in this part, which also clarifies its complicated composition and range of causes.

Fig. 1.1 Aerosols, their sources and composition

1.2.1 Natural Sources of Aerosols

Aerosols are released into the atmosphere by natural events such as dust storms, wildfires, volcanic eruptions, and biogenic emissions (Fig. 1.1). Massive amounts of mineral dust particles are produced by dust storms, which are caused by wind erosion of dry and semi-arid terrain (Qian et al. 2011). Long-distance transport by these particles can have an impact on visibility and air quality in areas distant from their source (Nair et al. 2005). The atmospheric composition and regional climate patterns are impacted by the emissions of smoke particles, organic carbon, and other aerosols caused by wildfires (Talukdar et al. 2021). Volcanic eruptions release ash, sulphur dioxide, and other volcanic aerosols into the atmosphere, which change air circulation and scatter sunlight, so affecting the global climate (Langmann 2014).

1.2.2 Anthropogenic Sources of Aerosols

Aerosol pollution is largely caused by human activity in several ways, such as energy generation, transportation, industrial processes, and agricultural practices (Haywood 2021) (Fig. 1.1). Industrial aerosols are created when pollutants like sulphur dioxide (SO_2), nitrogen oxides (NO_x), and particulate matter (PM) are released during activities including mining, manufacturing, and burning (He et al. 2014). These aerosols, which contain sulphate, nitrate, and black carbon particles, are responsible for the development of smog, acid rain, and respiratory ailments (Speight 2017). Transportation-related activities create exhaust emissions that contain aerosol particles and greenhouse gases, especially when fossil fuels are burned in cars, aeroplanes, and ships (Aakko-Saksa et al. 2023; Rönkkö et al. 2023). Vehicle exhaust produces fine particulate matter ($PM_{2.5}$), which is a component of urban air pollution that has an impact on ecosystems and human health (Board 2014). Sulphur dioxide and particulate matter are also released during the power-generating process in coal-fired power plants, which adds to local air pollution and atmospheric deposition. However, some aerosols and greenhouse gases are released into the atmosphere because of agricultural activities such as burning biomass, managing crop residue, and raising animals (Lima et al. 2022). Burning biomass releases organic carbon, smoke particles, and trace gases that change the composition of the atmosphere and local air quality. Ammonia emissions from cattle farming can combine with other contaminants to create secondary aerosols like ammonium nitrate (Wyer et al. 2022).

1.2.3 Composition of Aerosols

The variety of sources and atmospheric processes that contribute to the creation of aerosols is reflected in their compositions (Pöschl 2005). Apart from the primary aerosols that are released into the atmosphere directly, precursor gases such as ammonia, nitrogen oxides, sulphur dioxide, and volatile organic compounds also play a role in the formation of secondary aerosols (SOAs) (Tomasi and Lupi 2017). Secondary organic aerosols, nitrate aerosols, organic aerosols, and sulphate aerosols are all produced by these reactions and add to a load of anthropogenic and natural aerosols (Srivastava et al. 2022). Aerosol compositions show the diversity of atmospheric processes and sources that contribute to their formation. Sulphate aerosols are a significant element of fine particulate matter and play a role in acid deposition, aerosol pollution, and climate forcing. Sulphate aerosols are produced when sulphur dioxide (SO_2) emissions oxidise (Hirdman et al. 2010). Nitrogen oxides (NO_x) and ammonia (NH_3) emissions oxidise to produce nitrate aerosols, which are involved in atmospheric chemistry and regional air quality (Pai et al. 2021). Complex mixes of carbonaceous molecules released from both natural and man-made sources, such as burning biomass, burning fossil fuels, and biogenic emissions, make up organic aerosols (Qian et al. 2011). Black carbon absorbs sunlight and contributes to global warming and localized air pollution (Ajay et al. 2021). It is a part of the particulate matter released by incomplete combustion of fossil fuels, biomass, and biofuels. Mineral dust aerosols are produced as rocks and soil mechanically weather (Nair et al. 2005). They are common in dry and semi-arid areas and add to atmospheric dust loading, which has an impact on ecosystem dynamics, air quality, and temperature (Ajay et al. 2021).

1.3 Impacts of Aerosol Pollution

The Earth's climate system, atmospheric processes, air quality, and public health are all significantly impacted by aerosol pollution (Zhou et al. 2017; Li et al. 2017). These effects can be divided into two categories: aerosol-sun radiation interactions, or direct effects, and aerosol-cloud interactions, or indirect effects. Aerosols also contribute to the deterioration of air quality, which has a negative impact on health and increases the risk of early death, cardiovascular problems, and respiratory ailments.

1.3.1 Direct Effects of Aerosol Pollution

1.3.1.1 Scattering of Solar Radiation

Aerosols change the way solar energy is distributed in the atmosphere by diffracting incoming solar radiation both directly through reflection and indirectly through diffraction (Table 1.1). The Earth's energy budget is impacted by this scattering mechanism, which alters radiative forcing and regional temperature patterns (Quaas and Gryspeerdt 2022). Aerosols with light colours, like sulphate particles, tend to scatter more solar radiation, which cools the Earth's surface (Zhang 2020).

Table 1.1 Direct and indirect effects of aerosol pollution on atmospheric processes

Direct effects	Impact
Scattering of solar radiation	Reduces incoming solar radiation, leading to cooling of the atmosphere and surface temperatures
Absorption of solar radiation	Warms the atmosphere locally, especially in regions with high concentrations of absorbing aerosols such as black carbon
Changes in radiative forcing	Aerosols contribute to both positive (warming) and negative (cooling) radiative forcing depending on their composition, concentration, and altitude in the atmosphere
Indirect Effects	Impact
Cloud condensation nuclei (CCN)	Aerosols act as CCN, influencing cloud droplet formation and cloud albedo
Cloud microphysics	Aerosols can alter cloud droplet size distribution, cloud optical properties, and cloud lifetime
Precipitation processes	Aerosols may enhance or suppress precipitation, depending on factors such as aerosol type, concentration, and cloud dynamics

1.3.1.2   Absorption of Solar Radiation

Certain aerosol particles absorb incoming solar radiation, especially black carbon, and some forms of organic carbon, which causes the atmosphere to warm locally (Li et al. 2022) (Table 1.1). In locations with significant concentrations of black carbon aerosols, for example, in densely populated regions such as the Indo-Gangetic Plains, eastern China region, this absorption can cause the formation of urban heat islands and contribute to regional climate patterns (Wu et al. 2017).

1.3.2   Indirect Effects of Aerosol Pollution

1.3.2.1   Aerosol-Cloud Interactions

Aerosols provide surfaces for water vapour condensation and ice crystal production within clouds by acting as cloud condensation nuclei (CCN) and ice nuclei (IN) (Rosenfeld et al. 2016). Cloud longevity, precipitation processes, and droplet size distribution are all impacted by this process, which modifies cloud microphysics. Higher aerosol concentrations could produce higher cloud droplet concentrations, which may limit rainfall by causing smaller cloud droplets to form (Barthlott et al. 2022).

1.3.2.2   Modification of Cloud Properties

Cloud characteristics including albedo, optical thickness, and cloud fraction can also be influenced by aerosols (Barthlott et al. 2022). Cloud albedo and reflectivity can be increased by aerosol particles acting as CCN, which can result in the creation of more numerous but smaller cloud droplets (Quaas and Gryspeerdt 2022). The Earth's surface may cool because of this increase in cloud albedo reflecting more solar radiation back into space.

1.4   Impacts on Air Quality and Public Health

1.4.1   Respiratory Diseases

Breathing in fine particulate matter ($PM_{2.5}$) from aerosol pollution can irritate and inflame the respiratory system, making respiratory disorders including bronchitis, asthma, and chronic obstructive pulmonary disease worse (COPD) (Table 1.2) (Murray et al. 2020; Southerland et al. 2022). Lung cancer, decreased lung function, and respiratory infections have all been related to long-term high $PM_{2.5}$ exposure (Turner et al. 2011).

Table 1.2 Impacts of aerosol pollution on human health

Impacts	Consequences	Case study
Respiratory diseases	$PM_{2.5}$ can exacerbate respiratory conditions such as asthma, bronchitis, and chronic obstructive pulmonary disease (COPD)	Studies have shown that each $10 \, \mu g/m^3$ increase in $PM_{2.5}$ concentration is associated with a 4% increase in hospital admissions and 10–17% increase in mortality for respiratory diseases (Xing et al. 2016)
Cardiovascular disorders	Long-term exposure to $PM_{2.5}$ is associated with an increased risk of cardiovascular diseases, including heart attacks, strokes, and hypertension	Research indicates that long-term exposure to elevated PM2.5 levels is linked to a 12–14% higher risk of cardiovascular mortality (Jalali et al. 2021)
Premature mortality	Air pollution-related illnesses contribute to premature deaths, particularly among vulnerable populations such as children, the elderly, and individuals with pre-existing health conditions	Globally, air pollution is estimated to cause over 5 million premature deaths annually, with PM2.5 being a significant contributor (Thangavel et al. 2022)

1.4.2 Cardiovascular Disorders

An increased risk of cardiovascular conditions such as hypertension, heart attacks, strokes, and arrhythmias are linked to aerosol pollution (Lee et al. 2014) (Table 1.2). Through the lungs, fine particulate matter can enter the bloodstream and cause oxidative stress, endothelial dysfunction, and systemic inflammation (Jalali et al. 2021). These factors all have a role in the onset and progression of cardiovascular illnesses.

1.4.3 Premature Mortality

An increased risk of premature death is associated with long-term exposure to high aerosol pollution levels, especially fine particulate matter (Park et al. 2020). Aerosol pollution affects more than just respiratory and cardiovascular illnesses; it also affects other organ systems and increases the risk of premature death, particularly in vulnerable groups like the elderly, children, and people with underlying medical disorders when expose to longer periods such as months or years (Thangavel et al. 2022).

1.5 Mitigation Strategies for Aerosol Pollution

Public health, economic development, and environmental sustainability are all seriously threatened by aerosol pollution. Aerosol pollution must be addressed with a multimodal strategy that includes legislative actions, technology advancements, and behavioural adjustments (Fig. 1.2). This section examines the many mitigation techniques used throughout the world to lower aerosol emissions and enhance air quality, demonstrating their efficacy with case studies and real datasets.

1.5.1 Regulatory Measures: Clean Air Act in the United States

One of the most extensive legislative frameworks for managing air quality in the US is the Clean Air Act (CAA) ("Summary of the Clean Air Act I US EPA,"). The Clean Air Act (CAA), which was passed in 1970 and later updated, sets national air quality requirements for specific pollutants such as lead, carbon monoxide (CO), sulphur dioxide (SO_2), nitrogen oxides (NO_x), and particle matter ($PM_{2.5}$ and PM_{10}) (Hopke 2009). By means of emissions regulations, pollution controls, and permit requirements, the CAA gives the Environmental Protection Agency (EPA) the authority to limit emissions from automobiles, power plants, and other sources (McCarthy et al. 2014). The CAA has significantly improved air quality and public health outcomes

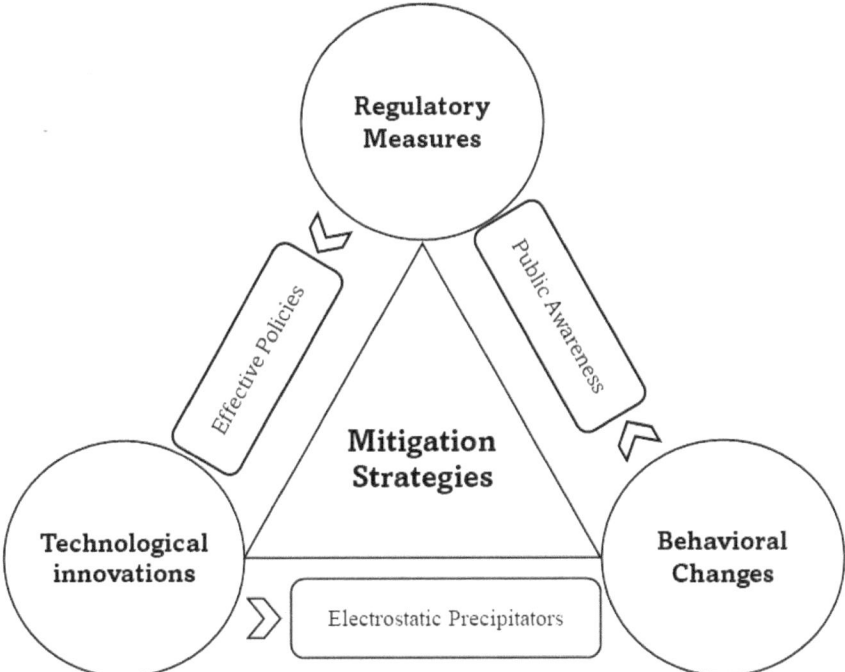

Fig. 1.2 Mitigation strategies for reducing air pollution

by lowering aerosol pollution levels across the country over the years. Emissions of major air pollutants covered by the Clean Air Act have significantly decreased, according to EPA data. The Clean Air Act regulates some air pollutants, and from 1970–2019, emissions of these pollutants have significantly decreased, according to EPA data. For instance, emissions of particulate matter ($PM_{2.5}$) dropped by 43%, sulphur dioxide (SO_2) by 91%, and nitrogen oxides (NO_x) by 74% between 1970 and 2019 (Ross et al. 2012). These decreases show how successful regulatory actions have been in reducing aerosol pollution and raising national air quality.

1.5.2 Technological Innovations: Electrostatic Precipitators in Industrial Settings

Particulate matter is extracted from industrial exhaust streams using electrostatic precipitators (ESPs), a type of air pollution control technology (Vallero 2024). Electrostatically charged aerosol particles are attracted to electrodes or collection plates via electrostatic attraction, which is how ESPs work (Miller 2005). To lower emissions of fine particulate matter, such as fly ash and black carbon, this technique is commonly used in steel mills, cement kilns, coal-fired power stations, and other industrial facilities (Kim et al. 2012). Research has indicated that electrostatic precipitators (ESPs) can effectively remove aerosol contaminants (up to 91.5%) with high removal efficiency, hence reducing air pollution and its associated environmental implications (Zhou et al. 2022). Electrostatic precipitators placed in coal-fired power stations can achieve particulate matter removal efficiency surpassing 99% (White 2013). Significant reductions in fine particulate matter emissions are achieved with this high degree of control, improving local populations' public health and air quality.

1.5.3 Behavioural Changes: Public Awareness Campaigns on Clean Cooking Practices

People are exposed to high concentrations of aerosol pollutants in many developing nations because domestic cooking and heating using solid fuels such as biomass, coal, and wood contributes to indoor and outdoor air pollution (Smith and Pillarisetti 2017). Household emissions of aerosols and other pollutants can be decreased with the support of public awareness campaigns and behavioural interventions that promote clean cooking methods including the use of clean fuels (such as liquefied petroleum gas) and upgraded cookstoves. To encourage the use of cleaner technology, these activities frequently include community engagement, educational initiatives, and financial incentives. WHO discovered that replacing conventional solid fuel cookstoves with cleaner models might significantly lower indoor air pollution levels and the health concerns that go along with them (WHO 2014). Adoption of better cookstoves, for instance, has been demonstrated to reduce carbon monoxide emissions by up to 50% and particulate matter emissions by up to 90%, improving respiratory health outcomes and lowering the prevalence of respiratory disorders in women and children (Phillip et al. 2023).

1.6 Aerosol Atmospheric Rivers (AARs): Formation and Dynamics

Aerosol atmospheric rivers (AARs) are elongated and narrow corridors of concentrated aerosol particles traversing the atmosphere over long distances (Chakraborty et al. 2022; Rautela et al. 2024a). The AARs shows similar characterises as Atmospheric Rivers (ARs) in shape but their transportation patterns are different (Singh and Goyal 2023b). Since AARs carry a dense payload of aerosol particles, influencing atmospheric dynamics, weather patterns, and regional climate variability (Rautela et al. 2024b).

1.6.1 Formation Mechanisms

AARs typically originate from regions with intense aerosol sources, such as industrial zones, densely populated regions, biomass burning areas, and dust-prone regions (Fig. 1.3) (Chakraborty et al. 2022). The formation of AARs involves several key processes.

1.6.1.1 Aerosol Emission Sources

AARs often originate from regions with high aerosol emissions, including urban areas, industrial complexes, greater sea depths, and agricultural activities (Chakraborty et al. 2022). Aerosols emitted from these sources, such as black carbon, dust, organic carbon, sea salts, and sulphates contribute to the formation of AARs (Chakraborty et al. 2018).

1.6.1.2 Atmospheric Transport

Once emitted, aerosol particles are transported by atmospheric circulation patterns, including wind currents and boundary layer dynamics (Yang et al. 2018; Yu et al. 2021). Long-range transport mechanisms, such as atmospheric jets and cyclonic systems, can carry aerosols over thousands of kilometres, facilitating the formation of AARs (Takahashi et al. 2018; Gallo et al. 2023).

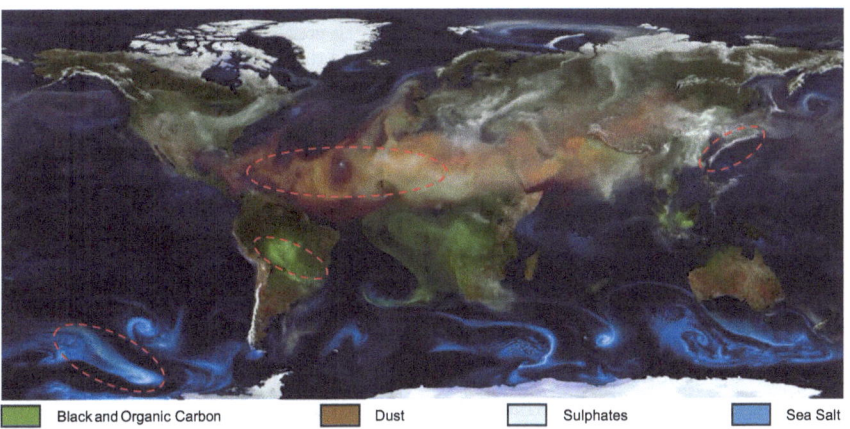

Black and Organic Carbon   Dust   Sulphates   Sea Salt

Fig. 1.3 Major Aerosols and their pathways over the globe

1.6.1.3   Concentration and Dispersion

As aerosols are transported through the atmosphere, they undergo processes of concentration and dispersion (von Schoenberg et al. 2021). Factors such as atmospheric stability, turbulence, and mixing processes influence the spatial distribution and concentration of aerosol particles within AARs (Chakraborty et al. 2018).

1.6.2   Dynamics of Aerosol Atmospheric Rivers

The dynamics of AARs are governed by a complex interplay of atmospheric processes, including synoptic-scale circulation patterns, boundary layer dynamics, and aerosol sources and sinks (Kulmala et al. 2023). Key aspects of AAR dynamics include:

1.6.2.1   Synoptic-Scale Circulation Patterns

AARs are often associated with large-scale weather systems, such as atmospheric ridges, troughs, and frontal boundaries (Michel et al. 2021). These synoptic-scale features provide the atmospheric dynamics necessary for the formation and propagation of AARs across different regions.

1.6.2.2   Boundary Layer Processes

The planetary boundary layer plays a crucial role in the vertical transport and dispersion of aerosol particles within AARs (Volná and Hladký 2020). Vertical mixing processes, including turbulence, convective currents, and stable stratification, influence the vertical distribution and residence time of aerosols within the boundary layer (Chakraborty et al. 2022).

1.6.2.3   Aerosol Sources and Sinks

Variations in aerosol sources and sinks influence the composition and characteristics of AARs. Anthropogenic sources, such as industrial emissions and vehicular exhaust, contribute to the formation of urban AARs, whereas natural sources, such as wildfires and dust storms, can generate regional-scale AARs with distinct aerosol compositions (Rautela et al. 2024a).

1.7 Conclusion

Air quality, human health, and climate are all significantly impacted by aerosol pollution, which is a complex environmental issue. The various aerosol sources and compositions highlight how complicated this problem is, calling for in-depth knowledge and practical mitigation measures. Even while laws like the US Clean Air Act have been successful in lowering aerosol emissions, technological advancements like electrostatic precipitators and alterations in behaviour like encouraging clean cooking methods are essential for reducing aerosol pollution even more. Furthermore, the appearance of AARs offers a fresh perspective on aerosol dynamics, with ramifications for weather patterns, atmospheric processes, and regional climate variability. Creating predictive tools and focused mitigation techniques require an understanding of the dynamics and formation mechanisms of AARs. To improve predictive accuracy and our capacity to predict the effects of AARs on atmospheric dynamics and air quality, future research in this field should concentrate on advanced detection techniques for AARs. For prediction of these, AARs should make use of machine learning algorithms and numerical weather prediction models. Efforts to mitigate the harmful effects of aerosol pollution and advocate for a cleaner, healthier environment for both current and future generations can be advanced by bridging these knowledge gaps and implementing innovative solutions.

References

Aakko-Saksa PT, Lehtoranta K, Kuittinen N et al (2023) Reduction in greenhouse gas and other emissions from ship engines: current trends and future options. Prog Energy Combust Sci 94:101055. https://doi.org/10.1016/j.pecs.2022.101055

Adebiyi A, Kok JF, Murray BJ et al (2023) A review of coarse mineral dust in the Earth system. Aeolian Res 60:100849. https://doi.org/10.1016/j.aeolia.2022.100849

Ajay A, Krishna Moorthy K, Satheesh SK, Ilavazhagan G (2021) Impact assessment of change in anthropogenic emissions due to lockdown on aerosol characteristics in a rural location. Curr Sci 120:332–340. https://doi.org/10.18520/cs/v120/i2/332-340

Amann M, Kiesewetter G, Schöpp W et al (2020) Reducing global air pollution: the scope for further policy interventions. Philos Trans R Soc A Math Phys Eng Sci 378:20190331. https://doi.org/10.1098/rsta.2019.0331

Aminzadegan S, Shahriari M, Mehranfar F, Abramović B (2022) Factors affecting the emission of pollutants in different types of transportation: a literature review. Energy Rep 8:2508–2529. https://doi.org/10.1016/j.egyr.2022.01.161

Barthlott C, Zarboo A, Matsunobu T, Keil C (2022) Importance of aerosols and shape of the cloud droplet size distribution for convective clouds and precipitation. Atmos Chem Phys 22:2153–2172. https://doi.org/10.5194/acp-22-2153-2022

Board CPC (2014) National air quality index. Cent Pollut Control Board, 1–58

Burns J, Boogaard H, Polus S, et al (2019) Interventions to reduce ambient particulate matter air pollution and their effect on health. Cochrane Database Syst Rev.https://doi.org/10.1002/14651858.CD010919.pub2

Chakraborty S, Fu R, Massie ST, Stephens G (2016) Relative influence of meteorological conditions and aerosols on the lifetime of mesoscale convective systems. Proc Natl Acad Sci 113:7426–7431. https://doi.org/10.1073/pnas.1601935113

Chakraborty S, Guan B, Waliser DE, et al (2021) Extending the atmospheric river concept to aerosols: climate and air quality impacts. Geophys Res Lett 48.https://doi.org/10.1029/2020GL091827

Chakraborty S, Guan B, Waliser DE, da Silva AM (2022) Aerosol atmospheric rivers: climatology, event characteristics, and detection algorithm sensitivities. Atmos Chem Phys 22:8175–8195. https://doi.org/10.5194/acp-22-8175-2022

Chakraborty S, Schiro KA, Fu R, Neelin JD (2018) On the role of aerosols, humidity, and vertical wind shear in the transition of shallow-to-deep convection at the Green Ocean Amazon 2014/5 site. Atmos Chem Phys 18:11135–11148. https://doi.org/10.5194/acp-18-11135-2018

Chen T, Liu Y, Ma Q et al (2019) Significant source of secondary aerosol: formation from gasoline evaporative emissions in the presence of SO_2 and NH_3. Atmos Chem Phys 19:8063–8081. https://doi.org/10.5194/acp-19-8063-2019

Gallo F, Uin J, Sanchez KJ et al (2023) Long-range transported continental aerosol in the eastern North Atlantic: three multiday event regimes influence cloud condensation nuclei. Atmos Chem Phys 23:4221–4246. https://doi.org/10.5194/acp-23-4221-2023

Goyal MK, Burn DH, Ojha CSP (2012) Evaluation of machine learning tools as a statistical down-scaling tool: temperatures projections for multi-stations for Thames River Basin, Canada. Theor Appl Climatol 108:519–534. https://doi.org/10.1007/s00704-011-0546-1

Goyal MK, Ojha CSP (2011) Evaluation of linear regression methods as downscaling tools in temperature projections over the Pichola Lake Basin in India. Hydrol Process 25:1453–1465. https://doi.org/10.1002/hyp.7911

Haywood J (2021) Atmospheric aerosols and their role in climate change. In: Climate change. Elsevier, pp 645–659

He H, Wang Y, Ma Q et al (2014) Mineral dust and NOx promote the conversion of SO_2 to sulfate in heavy pollution days. Sci Rep 4:4172. https://doi.org/10.1038/srep04172

Hidy GM (2019) Atmospheric aerosols: some highlights and highlighters, 1950 to 2018. Aerosol Sci Eng 3:1–20. https://doi.org/10.1007/s41810-019-00039-0

Hirdman D, Burkhart JF, Sodemann H et al (2010) Long-term trends of black carbon and sulphate aerosol in the Arctic: changes in atmospheric transport and source region emissions. Atmos Chem Phys 10:9351–9368. https://doi.org/10.5194/acp-10-9351-2010

Hopke PK (2009) Environmental chemometrics. In: Comprehensive chemometrics. Elsevier, pp 55–74

Jalali S, Karbakhsh M, Momeni M, et al (2021) Long-term exposure to PM2.5 and cardiovascular disease incidence and mortality in an Eastern Mediterranean country: findings based on a 15-year cohort study. Environ Heal 20:112. https://doi.org/10.1186/s12940-021-00797-w

Jha S, Das J, Goyal MK (2019) Assessment of risk and resilience of terrestrial ecosystem productivity under the influence of extreme climatic conditions over India. Sci Rep 9:18923. https://doi.org/10.1038/s41598-019-55067-0

Kim J-H, Yoo H-J, Hwang Y-S, Kim H-G (2012) Removal of particulate matter in a tubular wet electrostatic precipitator using a water collection electrode. Sci World J 2012:1–6. https://doi.org/10.1100/2012/532354

Koplitz SN, Mickley LJ, Marlier ME et al (2016) Public health impacts of the severe haze in Equatorial Asia in September–October 2015: demonstration of a new framework for informing fire management strategies to reduce downwind smoke exposure. Environ Res Lett 11:094023. https://doi.org/10.1088/1748-9326/11/9/094023

Kulmala M, Kokkonen T, Ezhova E et al (2023) Aerosols, clusters, greenhouse gases, trace gases and boundary-layer dynamics: on feedbacks and interactions. Boundary-Layer Meteorol 186:475–503. https://doi.org/10.1007/s10546-022-00769-8

Kumar N, Kumar Goyal M, Kumar Gupta A et al (2021a) Joint behaviour of climate extremes across India: past and future. J Hydrol 597:126185. https://doi.org/10.1016/j.jhydrol.2021.126185

Kumar N, Poonia V, Gupta BB, Goyal MK (2021b) A novel framework for risk assessment and resilience of critical infrastructure towards climate change. Technol Forecast Soc Change 165:120532. https://doi.org/10.1016/j.techfore.2020.120532

Kuniyal JC, Guleria RP (2019) The current state of aerosol-radiation interactions: a mini review. J Aerosol Sci 130:45–54. https://doi.org/10.1016/j.jaerosci.2018.12.010

Kuniyal JC, Kanwar N, Bhoj AS, et al (2021) Climate change impacts on glacier-fed and non-glacier-fed ecosystems of the Indian Himalayan Region : people ' s perception and adaptive strategies, 120

Langmann B (2014) On the role of climate forcing by volcanic sulphate and volcanic ash. Adv Meteorol 2014:1–17. https://doi.org/10.1155/2014/340123

Lee B-J, Kim B, Lee K (2014) Air pollution exposure and cardiovascular disease. Toxicol Res 30:71–75. https://doi.org/10.5487/TR.2014.30.2.071

Li J, Carlson BE, Yung YL et al (2022) Scattering and absorbing aerosols in the climate system. Nat Rev Earth Environ 3:363–379. https://doi.org/10.1038/s43017-022-00296-7

Li Z, Guo J, Ding A et al (2017) Aerosol and boundary-layer interactions and impact on air quality. Natl Sci Rev 4:810–833. https://doi.org/10.1093/nsr/nwx117

Lima MJA de, Nunes HGGC, Sampaio LS, et al (2022) Optimal soybean sowing window adjusted to climatic variability based on El Nino-Southern Oscillation using agrometeorological modeling. Pesqui Agropecuária Trop 52.https://doi.org/10.1590/1983-40632022v5272428

McCarthy JE, Copeland C, Parker L, Schierow LJ (2014) Clean Air Act: A summary of the act and its major requirements. New Trends Environ Sci, 147–168

Michel C, Sorteberg A, Eckhardt S et al (2021) Characterization of the atmospheric environment during extreme precipitation events associated with atmospheric rivers in Norway—Seasonal and regional aspects. Weather Clim Extrem 34:100370. https://doi.org/10.1016/j.wace.2021.100370

Miller BG (2005) Emissions Control Strategies for Power Plants. In: Coal energy systems. Elsevier, pp 283–392

Murray CJL, Aravkin AY, Zheng P et al (2020) Global burden of 87 risk factors in 204 countries and territories, 1990–2019: a systematic analysis for the Global Burden of Disease Study 2019. Lancet 396:1223–1249. https://doi.org/10.1016/S0140-6736(20)30752-2

Nair SK, Parameswaran K, Rajeev K (2005) Seven year satellite observations of the mean structures and variabilities in the regional aerosol distribution over the oceanic areas around the Indian subcontinent. Ann Geophys 23:2011–2030. https://doi.org/10.5194/angeo-23-2011-2005

Pai SJ, Heald CL, Murphy JG (2021) Exploring the global importance of atmospheric ammonia oxidation. ACS Earth Sp Chem 5:1674–1685. https://doi.org/10.1021/acsearthspacechem.1c00021

Park S, Allen RJ, Lim CH (2020) A likely increase in fine particulate matter and premature mortality under future climate change. Air Qual Atmos Heal 13:143–151. https://doi.org/10.1007/s11869-019-00785-7

Phillip E, Langevin J, Davis M et al (2023) Improved cookstoves to reduce household air pollution exposure in sub-Saharan Africa: a scoping review of intervention studies. PLoS ONE 18:e0284908. https://doi.org/10.1371/journal.pone.0284908

Poonia V, Goyal MK, Gupta BB et al (2021) Drought occurrence in different river basins of India and blockchain technology based framework for disaster management. J Clean Prod 312:127737. https://doi.org/10.1016/j.jclepro.2021.127737

Pöschl U (2005) Atmospheric aerosols: composition, transformation, climate and health effects. Angew Chemie Int Ed 44:7520–7540. https://doi.org/10.1002/anie.200501122

Pryor SC, Crippa P, Sullivan RC (2015) Atmospheric chemistry. In: Reference Module in earth systems and environmental sciences. Elsevier

Qian Y, Flanner MG, Leung LR, Wang W (2011) Sensitivity studies on the impacts of Tibetan Plateau snowpack pollution on the Asian hydrological cycle and monsoon climate. Atmos Chem Phys 11:1929–1948. https://doi.org/10.5194/acp-11-1929-2011

Quaas J, Gryspeerdt E (2022) Aerosol-cloud interactions in liquid clouds. In: Aerosols and climate. Elsevier, pp 489–544

Rautela KS, Singh S, Goyal MK (2024a) Characterizing the spatio-temporal distribution, detection, and prediction of aerosol atmospheric rivers on a global scale. J Environ Manage 351:119675. https://doi.org/10.1016/j.jenvman.2023.119675

Rautela KS, Singh S, Goyal MK (2024b) Resilience to air pollution: A novel approach for detecting and predicting aerosol atmospheric rivers within earth system boundaries earth systems and environment. https://doi.org/10.1007/s41748-024-00421-0

Rönkkö T, Saarikoski S, Kuittinen N et al (2023) Review of black carbon emission factors from different anthropogenic sources. Environ Res Lett 18:033004. https://doi.org/10.1088/1748-9326/acbb1b

Rosenfeld D, Zheng Y, Hashimshoni E et al (2016) Satellite retrieval of cloud condensation nuclei concentrations by using clouds as CCN chambers. Proc Natl Acad Sci 113:5828–5834. https://doi.org/10.1073/pnas.1514044113

Ross K, Chmiel JF, Ferkol T (2012) The impact of the clean air act. J Pediatr 161:781–786. https://doi.org/10.1016/j.jpeds.2012.06.064

Sarkar S, Chauhan A, Kumar R, Singh RP (2019) Impact of deadly dust storms (May 2018) on air quality, meteorological, and atmospheric parameters over the northern parts of India. GeoHealth 3:67–80. https://doi.org/10.1029/2018GH000170

Seinfeld JH (2015) Tropospheric chemistry and composition I aerosols/particles. In: Encyclopedia of atmospheric sciences. Elsevier, pp 182–187

Sharma A, Goyal MK (2020) Assessment of the changes in precipitation and temperature in Teesta River basin in Indian Himalayan Region under climate change. Atmos Res 231:104670. https://doi.org/10.1016/j.atmosres.2019.104670

Silva AM, Maring H, Seidel F, et al (2020) Aerosol , cloud , ecosystems (ACE) final study report

Singh S, Goyal MK (2023a) An innovative approach to predict atmospheric rivers: exploring convolutional autoencoder. Atmos Res 289:106754. https://doi.org/10.1016/j.atmosres.2023.106754

Singh S, Goyal MK (2023b) Enhancing climate resilience in businesses: the role of artificial intelligence. J Clean Prod 418:138228. https://doi.org/10.1016/j.jclepro.2023.138228

Singh S, Goyal MK, Jha S (2023) Role of large-scale climate oscillations in precipitation extremes associated with atmospheric rivers: nonstationary framework. Hydrol Sci J 68:395–411. https://doi.org/10.1080/02626667.2022.2159412

Smith KR, Pillarisetti A (2017) Household air pollution from solid cookfuels and its effects on health. Dis control priorities, Third Ed (Volume 7) Inj Prev Environ Heal, 133–152. https://doi.org/10.1596/978-1-4648-0522-6_CH7

Southerland VA, Brauer M, Mohegh A et al (2022) Global urban temporal trends in fine particulate matter (PM2·5) and attributable health burdens: estimates from global datasets. Lancet Planet Heal 6:e139–e146. https://doi.org/10.1016/S2542-5196(21)00350-8

Speight JG (2017) Sources and types of organic pollutants. In: Environmental organic chemistry for engineers. Elsevier, pp 153–201

Srivastava D, Vu T V., Tong S, et al (2022) Formation of secondary organic aerosols from anthropogenic precursors in laboratory studies. npj Clim Atmos Sci 5:22. https://doi.org/10.1038/s41612-022-00238-6

Summary of the Clean Air Act I US EPA. https://www.epa.gov/laws-regulations/summary-clean-air-act. Accessed 4 Mar 2024

Takahashi HG, Watanabe S, Nakata M, Takemura T (2018) Response of the atmospheric hydrological cycle over the tropical Asian monsoon regions to anthropogenic aerosols and its seasonality. Prog Earth Planet Sci 5:44. https://doi.org/10.1186/s40645-018-0197-2

Talukdar S, Tripathi SN, Lalchandani V et al (2021) Air pollution in New Delhi during late winter: an overview of a group of campaign studies focusing on composition and sources. Atmosphere (basel) 12:1432. https://doi.org/10.3390/atmos12111432

Thangavel P, Park D, Lee Y-C (2022) Recent insights into particulate matter (PM2.5)-mediated toxicity in humans: an overview. Int J Environ Res Public Health 19:7511. https://doi.org/10.3390/ijerph19127511

Tomasi C, Lupi A (2017) Primary and secondary sources of atmospheric aerosol. In: Atmospheric aerosols. Wiley, pp 1–86

Turner MC, Krewski D, Pope CA et al (2011) Long-term ambient fine particulate matter air pollution and lung cancer in a large cohort of never-smokers. Am J Respir Crit Care Med 184:1374–1381. https://doi.org/10.1164/rccm.201106-1011OC

Vallero DA (2024) Air pollution control technologies. In: Air Pollution calculations. Elsevier, pp 441–497

Volná V, Hladký D (2020) Detailed assessment of the effects of meteorological conditions on PM10 concentrations in the Northeastern part of the Czech Republic. Atmosphere (basel) 11:497. https://doi.org/10.3390/atmos11050497

von Schoenberg P, Tunved P, Grahn H et al (2021) Aerosol dynamics and dispersion of radioactive particles. Atmos Chem Phys 21:5173–5193. https://doi.org/10.5194/acp-21-5173-2021

Wang J, Strong K (2019) British Columbia's forest fires, 2018. Stat Canada 8

White HJ (2013) Role of electrostatic precipitators in particulate control: a retrospective and prospective view. J Air Pollut Control Assoc 25:102–107. https://doi.org/10.1080/00022470.1975.10470052

WHO (2014) World health statistics

Wu H, Wang T, Riemer N et al (2017) Urban heat island impacted by fine particles in Nanjing. China. Sci Rep 7:11422. https://doi.org/10.1038/s41598-017-11705-z

Wyer KE, Kelleghan DB, Blanes-Vidal V et al (2022) Ammonia emissions from agriculture and their contribution to fine particulate matter: a review of implications for human health. J Environ Manage 323:116285. https://doi.org/10.1016/j.jenvman.2022.116285

Xing Y-F, Xu Y-H, Shi M-H, Lian Y-X (2016) The impact of PM2.5 on the human respiratory system. J Thorac Dis 8:E69-74. https://doi.org/10.3978/j.issn.2072-1439.2016.01.19

Yang J, Kang S, Ji Z, Chen D (2018) Modeling the origin of anthropogenic black carbon and its climatic effect over the Tibetan plateau and surrounding regions. J Geophys Res Atmos 123:671–692. https://doi.org/10.1002/2017JD027282

Yu H, Tan Q, Zhou L et al (2021) Observation and modeling of the historic "Godzilla" African dust intrusion into the Caribbean Basin and the southern US in June 2020. Atmos Chem Phys 21:12359–12383. https://doi.org/10.5194/acp-21-12359-2021

Zhang B (2020) The effect of aerosols to climate change and society. J Geosci Environ Prot 08:55–78. https://doi.org/10.4236/gep.2020.88006

Zhou L, Zhang J, Liu X et al (2022) Improving the electrostatic precipitation removal efficiency on fine particles by adding wetting agent during the chemical agglomeration process. Fuel Process Technol 230:107202. https://doi.org/10.1016/j.fuproc.2022.107202

Zhou S, Collier S, Jaffe DA et al (2017) Regional influence of wildfires on aerosol chemistry in the western US and insights into atmospheric aging of biomass burning organic aerosol. Atmos Chem Phys 17:2477–2493. https://doi.org/10.5194/ACP-17-2477-2017

Chapter 2
Aerosol Atmospheric Rivers: Detection and Spatio-Temporal Patterns

Abstract Aerosol Atmospheric Rivers (AARs) are a complex environmental phenomena that have a significant impact on the sustainability of the environment and human health. The formation, dynamics, types, causes, health effects, monitoring, mitigation, adaptive techniques, future directions, and research requirements associated with AARs are all thoroughly examined in this study. Similar to traditional atmospheric rivers (ARs), AARs carry a wide variety of aerosols over great distances, such as dust, black carbon, dust, organic carbon, sea salt, and sulphates, influencing global environmental conditions and air quality. The monitoring of AARs utilizes datasets such as the MERRA-2 hourly aerosol reanalysis dataset, coupled with an advanced global AR algorithm to detect the shapes of AARs of key aerosol species. Supporting the characterization of aerosol composition, sources, transport modes, and spatio-temporal distribution, these initiatives contribute to the development of thorough evaluations of the health effects of air pollution caused by AAR. AAR-induced aerosol pollution has both acute and long-term health impacts, especially for vulnerable populations over the Indo-Gangetic Plains, Eastern China, South African regions, western USA, etc. This makes interdisciplinary cooperation and adaptive mitigation solutions necessary. Future research will concentrate on interdisciplinary collaboration, technical advancements, and long-term assessments of health consequences to address novel issues arising from AARs and safeguard public health against climate change. However, this study emphasizes how critical it is to understand and reduce the health concerns linked to extreme aerosol pollution caused by AARs through cooperative research, creative technology, and proactive governmental actions.

Keywords Aerosol Atmospheric Rivers (AARs) · Health effects · Monitoring · Sustainability

© The Author(s), under exclusive license to Springer Nature Switzerland AG 2024   19
M. K. Goyal and K. S. Rautela, *Aerosol Atmospheric Rivers*,
SpringerBriefs in Applied Sciences and Technology,
https://doi.org/10.1007/978-3-031-66758-9_2

2.1 Introduction

Natural rivers play a crucial part in the hydrological cycle of Earth, serving as key watercourses that shape landscapes and support ecosystems (Kumar et al. 2020; Sofi et al. 2021; Rautela et al. 2024a). Similarly, the shape of atmospheric rivers (ARs) resembles that of the natural rivers but carries a tremendous volume of water in through the atmosphere (Singh et al. 2023a). Traditional ARs travel through the atmosphere, transferring moisture from tropical regions to higher latitudes, whereas natural rivers flow across the surface, propelled by gravity and impacted by precipitation, melting, and runoff (Sinha et al. 2019; Jha et al. 2019; Dubey and Goyal 2020; Bozkurt et al. 2021; Rautela et al. 2023a, b). The maintenance of life and ecosystems depends on both occurrences, which are essential to the distribution of water (Loucks and van Beek 2017). Traditional ARs contribute to the global water cycle by sustaining strong precipitation events like heavy rain and flooding upon landfall, whereas natural rivers directly supply water resources for human activity (Singh and Goyal 2023a). Natural processes have produced both, demonstrating how Earth's atmospheric dynamics and water systems are intertwined.

Similar to conventional ARs, aerosol atmospheric rivers (AARs) are characterized by the long-distance transportation of large concentrations of aerosols—microscopic particles suspended in the atmosphere (Chakraborty et al. 2021, 2022; Rautela et al. 2024b). AARs also carry aerosols, which have an impact on environmental health and regional as well as global air quality, in addition to the moisture that regular ARs primarily carry (Chakraborty et al. 2021). Both have long, thin forms that highlight their function as dynamic carriers of atmospheric elements. Because of its profound effects on respiratory health and air quality, research on AARs is critical to human health (Rautela et al. 2024b). The composition of the air that people breathe is directly impacted by these long, narrow passageways that carry high quantities of aerosols, such as pollutants and particulate matter, over great distances (Chakraborty et al. 2018). Aerosols in AARs raise the risk of cardiovascular illnesses and aggravate respiratory disorders like asthma and chronic obstructive pulmonary disease (COPD) (Jiang et al. 2016). To mitigate the detrimental health effects associated with aerosol exposure resulting from these atmospheric occurrences, it is critical to understand the dynamics of AARs and how they affect air quality (Chakraborty et al. 2021).

2.2 Formation and Dynamics of Aerosol Atmospheric Rivers

2.2.1 Factors Contributing to the Formation of AARs

AARs are complex phenomena that are impacted by numerous interconnected factors (Rautela et al. 2024b). Dominant wind patterns are one of these; they act as conduits for the long-range transport of aerosols, which causes concentrated streams to accumulate in the atmosphere (Lapere et al. 2024). One of the primary elements influencing the development of AARs is dense aerosol plumes that are directional due to strong trade winds (Chakraborty et al. 2021). Moreover, the surrounding geology greatly influences the behaviour of AARs (Li et al. 2022). Aerosol concentration and movement are influenced by topographical features such as mountain ranges and beaches, which act as airflow barriers or conduits. Seasonal fluctuations can alter the frequency and intensity of AARs due to altered air circulation patterns and the existence of sources of aerosol emissions, further complicating the dynamics of these events (Rautela et al. 2024b). These aerosols are the primary constituent of particulate matter such as $PM_{2.5}$ and PM_{10} that have a relatively greater impact on human health (Chakraborty et al. 2021). However, from the regional point of view, the air quality has declined recently due to several noteworthy extreme aerosol events, including the Australian bushfires in 2021 and the Indian dust storm in 2018 (Sarkar et al. 2019; Filkov et al. 2020). These extreme events have raised aerosol concentrations in the atmosphere and posed risks to human health, increase mortality, ecosystems, the economy, and meteorological phenomena (Rautela et al. 2024c). The incorporation of cutting-edge technologies, such as chemical transport models and satellite remote sensing, enhances our understanding of aerosol dynamics and makes it easier to develop mitigation methods and adaptations to changing climatic conditions (Tian et al. 2023). Important information about aerosol behaviour, evolution, and implications for weather patterns and climate variations are obtained by satellite observations, numerical modelling, and machine learning techniques (Yan et al. 2018; Shakya et al. 2023). The application of an interdisciplinary methodology enables the thorough observation and analysis of aerosol dynamics, a crucial step in addressing pressing environmental and public health concerns. For example, seasonal monsoon systems can have an impact on the distribution and transport of aerosols in regions that regularly see significant rainfall and atmospheric moisture levels (Lau et al. 2017). Furthermore, by altering temperature regimes, precipitation patterns, and air circulation dynamics, long-term changes in global climate patterns may exacerbate AARs. This gives rise to grave concerns regarding AARs' future. These changes may have a substantial effect on the behaviour and distribution of AARs, which could have consequences for air quality, climate stability, and environmental health. Understanding the complex interactions among various factors that influence the emergence and development of AARs is essential for appreciating their multifaceted influence and formulating workable strategies to alleviate their detrimental consequences on the environment (Gollakota et al. 2021).

2.2.2 Role of Aerosols in AAR Formation and Transport

Aerosols play a crucial role in the formation and transport of AARs, which are concentrated pathways similar to ARs (Singh and Goyal 2023b; Rautela et al. 2024b; Rautela et al. 2024c). These aerosols originate from both natural sources like forest fires, dust, and sea salt, as well as human-made sources such as industrial emissions and vehicular exhaust (Jacob et al. 1986; Christensen et al. 2022). The formation of AARs is influenced by atmospheric conditions including humidity, temperature, wind speed, and the vertical stability of aerosol species (Chakraborty et al. 2021). Aerosols act as nuclei around which water vapour can condense, forming cloud droplets. This process can lead to the creation of dense aerosol-laden clouds within AARs, which can travel long distances across continents (Karydis et al. 2011). Aerosol species such as black carbon, dust, organic carbon, sea salt, and sulfates contribute to the composition of AARs, with different aerosol types exhibiting varying transportation patterns and dispersion characteristics (Chakraborty et al. 2022). AARs can extend over vast distances, shaping air quality and environmental conditions across diverse regions. The concentration and disposition of aerosol particles within AARs can significantly impact visibility, weather patterns, and climate dynamics on both local and global scales (Che et al. 2024). Understanding the role of aerosols in AAR formation and transport is essential for assessing their impacts on environmental quality, human health, and ecosystem stability. Efforts to monitor and mitigate aerosol pollution can help mitigate the adverse effects associated with AARs and improve overall air quality and environmental health.

2.2.3 Interaction Between AARs and Meteorological Phenomena

The interaction between AARs and meteorological phenomena is a complex and dynamic process that significantly impacts weather patterns and atmospheric conditions. AARs, characterized by concentrated aerosol particles transported within the atmosphere, interact with various meteorological factors, such as humidity, temperature, and wind patterns (Wen et al. 2020). These interactions influence the formation, intensity, and trajectory of AARs, shaping their impact on regional and global climate systems. One crucial aspect of this interaction is the role of AARs in modulating precipitation patterns (Persad 2023). AARs can act as nuclei for cloud formation, affecting the microphysical properties of clouds and influencing precipitation efficiency (Rosenfeld et al. 2016). The presence of aerosol particles within AARs can alter cloud droplet size distribution, cloud lifetime, and precipitation intensity, leading to changes in rainfall patterns and distribution (Ramanathan et al. 2001; Yang et al. 2018). Furthermore, AARs can influence atmospheric dynamics, including atmospheric stability and circulation patterns (Chakraborty et al. 2022). The injection of aerosol particles into the atmosphere can modify temperature gradients and pressure systems, potentially affecting the development of weather systems such as cyclones, storms, and atmospheric fronts (Zhang 2020). Additionally, AARs can exacerbate extreme weather events by enhancing atmospheric instability and moisture convergence, leading to the intensification of storms and precipitation extremes (Persad 2023). Conversely, meteorological conditions, such as wind patterns and atmospheric circulation, can also influence the dispersion and concentration of aerosol particles within AARs, shaping their spatial distribution and impact on air quality and visibility (Singh et al., 2023b; Poonia et al. 2021; Chakraborty et al. 2022). Understanding these interactions is essential for improving climate modelling, weather forecasting, and mitigating the impacts of aerosol pollution on both local and global scales.

2.3 Aerosols Atmospheric Rivers (AARs): Types and Sources

The aerosols in AARs encompass a diverse range of particles, each with its own unique characteristics and origins (Rautela et al. 2024b). One prominent type of aerosol commonly transported by AARs is black carbon (BC) (Chakraborty et al. 2021). Originating predominantly from the combustion of fossil fuels, biomass burning, and industrial processes, black carbon particles are known for their light-absorbing properties, contributing significantly to atmospheric warming and climate change (Zhao et al. 2022). Dust particles, another significant component of aerosols carried by AARs, originate from natural sources such as desert regions, arid landscapes, and agricultural activities (Nair et al. 2005; Qian et al. 2011). These particles are comprised of mineral dust and soil components, often lifted into the atmosphere by wind erosion or anthropogenic activities (Lapere et al. 2024). Dust aerosols can have widespread impacts on air quality, visibility, and respiratory health, especially in regions prone to dust storms and desertification. Organic carbon (OC) aerosols, derived from both natural and anthropogenic sources, are also transported by AARs (Pöschl 2005; Chakraborty et al. 2021). Anthropogenic sources of organic carbon include burning biomass, industrial operations, and automobile emissions. Natural sources of organic carbon include biogenic emissions from vegetation and microbiological activities (Kashyap et al. 2019). Organic carbon aerosols play a significant role in atmospheric physical processes and chemistry, affecting air quality, cloud formation, and climate dynamics (Gupta et al. 2022). Sea salt aerosols represent another important category of particles transported by AARs, originating primarily from the ocean surface through processes such as wave breaking, sea spray, and bubble bursting (Ovadnevaite et al. 2014). These aerosols contain sodium chloride and other salts, with their production influenced by factors such as wind speed, sea surface temperature, and oceanic currents (Nair et al. 2005). Sea salt aerosols contribute to cloud formation, atmospheric chemistry, and marine biogeochemical cycles (Horowitz et al. 2020). Another type of aerosol, Sulfate aerosols, generated from both natural and anthropogenic sources, are prevalent in the atmosphere

Table 2.1 Residence time of different types of aerosols in the atmosphere

Aerosol type	Residence time range	Main factors influencing residence time
Black Carbon	Days to Weeks	Atmospheric dynamics, deposition, and removal processes
Dust	Days to Weeks	Particle size, wind speed, atmospheric stability
Organic Carbon	Hours to Days	Chemical reactions, atmospheric mixing
Sea salt	Hours to Days	Wind speed, humidity, sea surface conditions
Sulphates	Days to Weeks	Atmospheric chemistry, precipitation, deposition

and are frequently transported by AARs (Rasch et al. 2008). Natural sources of sulfate aerosols include volcanic eruptions and biogenic emissions, while anthropogenic sources encompass industrial activities, combustion processes, and agricultural practices (Artaxo et al. 2022). Sulfate aerosols play a crucial role in atmospheric chemistry, cloud formation, and radiative forcing, with implications for climate change and air quality regulation (Hirdman et al. 2010). However, residence times of different aerosol types vary widely, ranging from hours to weeks, depending on factors such as atmospheric dynamics, chemical reactions, and deposition processes (Table 2.1). Factors influencing residence times include atmospheric stability, wind speed, humidity, and sea surface conditions, which determine the transport and removal mechanisms for each aerosol type.

2.4   Health Impacts of Aerosol Atmospheric Rivers

AARs represent a significant environmental phenomenon with far-reaching consequences, not only for the planet's climate, but also for human health (Chakraborty et al. 2021). The inhalation hazards posed by different types of aerosols within AARs which are the prime constituents of $PM_{2.5}$ and PM_{10} can have both short-term and long-term health effects, particularly impacting vulnerable populations and regions (Gurjar et al. 2016; Oh et al. 2020; Nagpure and Lal 2022). Firstly, the composition of aerosols within AARs varies widely, encompassing particles such as black carbon, dust, organic carbon, sea salt, and sulfates (Rautela et al. 2024b). Each of these aerosol types presents unique inhalation hazards. For instance, black carbon particles, emitted from sources like vehicular exhaust and biomass burning, are fine particulate matter that can penetrate deep into the lungs upon inhalation, leading to respiratory issues and exacerbating conditions like asthma and bronchitis (Manisalidis et al. 2020). Dust aerosols, originating from deserts and dry regions, carry allergens and pathogens, posing risks of respiratory infections and allergic reactions (Fussell and Kelly 2021). Organic carbon aerosols, produced by biomass burning and industrial activities, contain volatile organic compounds that can irritate the respiratory system and contribute to the formation of smog (Chen et al. 2017). Sea salt aerosols, generated by ocean spray, may contain microbes and toxins, potentially causing respiratory infections and exacerbating existing respiratory conditions (Biddle et al. 2021). However, sulfate aerosols, arising from industrial emissions and volcanic eruptions, can irritate the respiratory tract and contribute to the formation of acid rain (Manisalidis et al. 2020).

Exposure to AARs can result in both short-term and long-term health effects. In the short term, individuals exposed to high concentrations of aerosols within AARs may experience symptoms such as coughing, wheezing, shortness of breath, and irritation of the eyes, nose, and throat (Salin et al. 2021). These acute respiratory symptoms can be particularly severe in vulnerable populations such as children, the elderly, individuals with pre-existing respiratory conditions, and those with compromised immune systems (Nagpure and Lal 2022). Moreover, short-term exposure to AARs has been linked to increased hospital admissions for respiratory illnesses and cardiovascular events, as well as higher mortality rates (Palacio et al. 2023). In the long term, chronic exposure to aerosols within AARs can have more serious health consequences. Prolonged inhalation of fine particulate matter, such as black carbon and sulfate aerosols, has been associated with the development and exacerbation of chronic respiratory diseases, including chronic obstructive pulmonary disease (COPD), emphysema, and lung cancer (Pope et al. 2004; Basith et al. 2022). Additionally, exposure to AARs has been linked to cardiovascular diseases, neurological disorders, and adverse birth outcomes (Lee et al. 2014). Furthermore, the deposition of aerosol particles in the respiratory system can trigger inflammatory responses, leading to the progression of respiratory diseases and the impairment of lung function over time.

Certain populations and regions are disproportionately affected by AAR-related health issues (Gurjar et al. 2016). Urban areas with high levels of industrial activity, vehicular traffic, and population density are particularly vulnerable to elevated concentrations of aerosols within AARs (Wu and Boor 2021). Low-income communities and marginalized populations living near industrial facilities and transportation hubs may face heightened health risks due to exposure to air pollutants from AARs (Rentschler and Leonova 2023). Furthermore, regions prone to frequent AAR events, such as densely industrial and populated areas and areas downwind of major pollution sources, experience persistent air quality problems and elevated rates of respiratory and cardiovascular diseases among residents (Lee et al. 2014).

2.5 Monitoring of Aerosol Atmospheric Rivers

2.5.1 Datasets

To monitor AARs effectively, the case study utilizes a variety of datasets to gather crucial information on aerosol transport, concentration, and distribution. Among the most employed datasets is the MERRA-2 hourly aerosol reanalysis dataset (Gelaro et al. 2017; Randles et al. 2017). This dataset provides comprehensive information on the zonal and meridional components of integrated aerosol transport (IAT) across the entire atmospheric column, spanning from 0.1 hPa to 1000 hPa. The MERRA-2 dataset offers a spatial resolution of $0.5° \times 0.625°$ and a temporal resolution of 1 h, enabling detailed analysis of aerosol dynamics over time and space. This study primarily focuses on key aerosol species such as black carbon (BC), dust (DU), organic carbon (OC), sea salt (SS), and sulfates (SU) extracted from the MERRA-2 dataset for their monitoring and analysis purposes.

2.5.2 Methodology

The monitoring of AARs involves the development and implementation of robust methodologies to detect and analyze these atmospheric phenomena accurately. This study made some enhancements in the global AR algorithms to identify AARs and further reconstruct past events (Ralph et al. 2020; Chakraborty et al. 2021, 2022). Initially, zonal and meridional components of IAT values are taken to compute the resultant IAT values (Fig. 2.1). Further, this algorithm involves computing extreme IAT events for each grid and time-step by determining the 85th percentile of resultant IAT for specific aerosol species (Fig. 2.2). Subsequent refinements in the methodology focus on coherence, filtering, and orientation criteria to ensure accurate detection of AARs (Rautela et al. 2024b).

2.5.3 Outcomes

The monitoring of AARs yields valuable outcomes that contribute to our understanding of aerosol pollution and its environmental impacts. By analysing AAR availability through the AARs algorithm shows the presence of AARs at the grid at a particular time step. A more than 100,000 + AARs are detected all over the globe during the period of 2015–2022 at 6 hourly time frame. Here we show only the AARs shape at an arbitrary time step in Fig. 2.3 to visualize the shape of AARs of key aerosols.

The distribution and intensity of different aerosol species in the atmosphere exhibit complex patterns that vary by season and region. Black carbon (BC) aerosol-induced air quality deterioration peaks in southern African regions and eastern China during summers but shifts to the Indo-Gangetic-Brahmaputra (IGB) plains and eastern China during winters, driven by biomass burning and anthropogenic activities (Ramachandran et al. 2020; Artaxo et al. 2022). Conversely, desert dust (DU) aerosols are concentrated over the northern side of the African continent and regions with dry conditions, originating from deserts like Sahara and Taklamakan, with higher frequency during summers due to anticyclonic motion over the Sahara Desert (Xu et al. 2020; Hu et al. 2022; Merdji et al. 2023). Organic carbon (OC) aerosols, sourced from

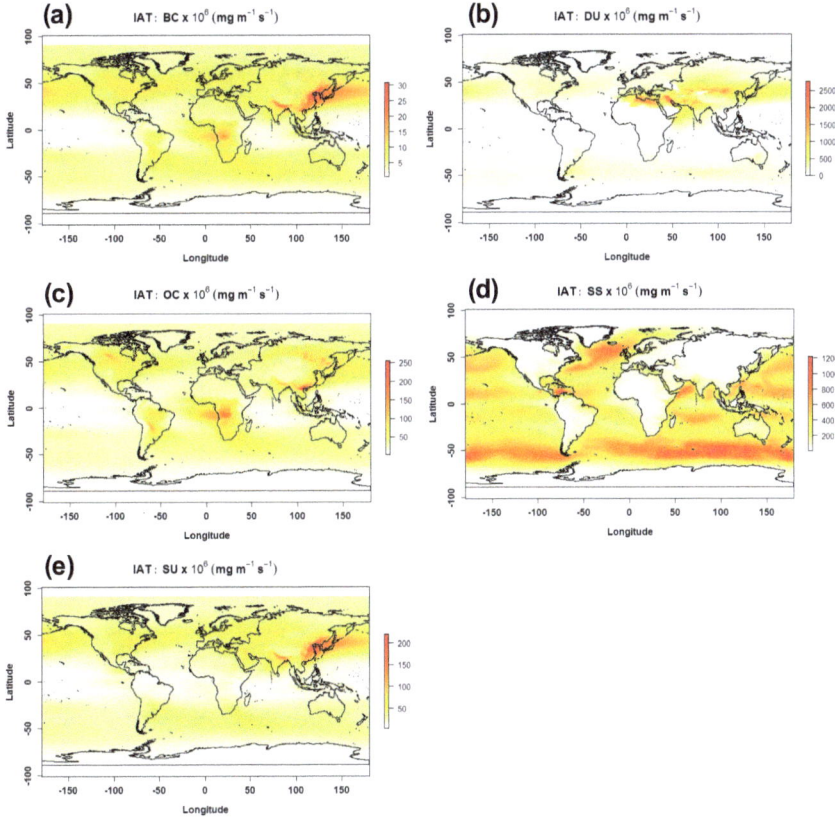

Fig. 2.1 IAT values of different aerosol species

boreal and tropical rainforests, are more pronounced in southern China and northern Russia during summers, whereas their intensity increases but frequency decreases in regions like the Congo and eastern India during winters (Artaxo et al. 2022; May et al. 2023). Sea salt (SS) aerosols exhibit an intensified presence in the southern hemisphere during summers and in mid-latitudes during winters, influenced by convective actions and synoptic storm activities (Chakraborty et al. 2021). Sulfate (SU) aerosols, notably intense over regions from China to the North Pacific Ocean, are influenced by anthropogenic emissions, with significant releases in the eastern USA and IGB plains, exhibiting consistent intensity throughout the year (Dai et al. 2019; Zhang et al. 2021). These aerosol dynamics underscore the complex interplay between natural processes, human activities, and atmospheric circulation patterns, highlighting the need for comprehensive monitoring and mitigation strategies to address their diverse health and environmental impacts. Moreover, the integration of advanced monitoring technologies such as satellite remote sensing, lidar, and ground-based stations enables

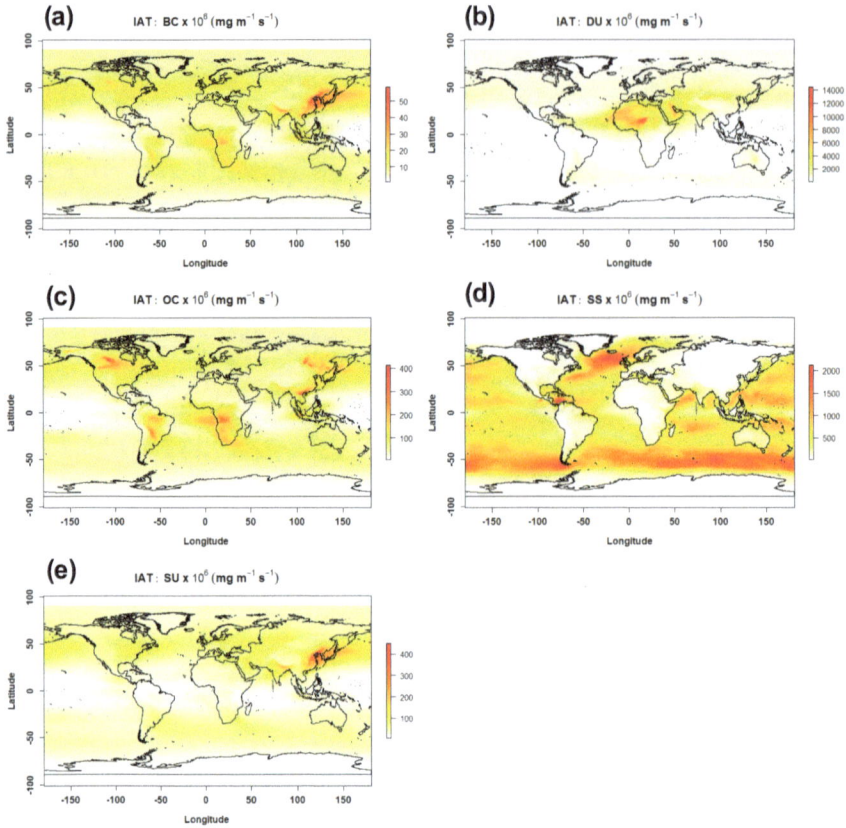

Fig. 2.2 Extreme aerosol transport values by taking the 85th percentile limit of different aerosol species

researchers to acquire high-resolution data on aerosol dynamics from diverse platforms, both globally and regionally (Filonchyk et al. 2019). These outcomes facilitate the development of efficient mitigation strategies and adaptation measures to address environmental challenges associated with aerosol pollution.

2.5.4 Reconstruction of Past Events and Checking the Availability of AARs

In reconstructing past events such as the significant dust storms and haze episodes that have afflicted regions like India and South Asia, it is crucial to assess the role of AARs in exacerbating these environmental, health, and economic crises (Figs. 2.4 and 2.5) (Table 2.2). In 2015, Equatorial Asia experienced large-scale forest fires, resulting

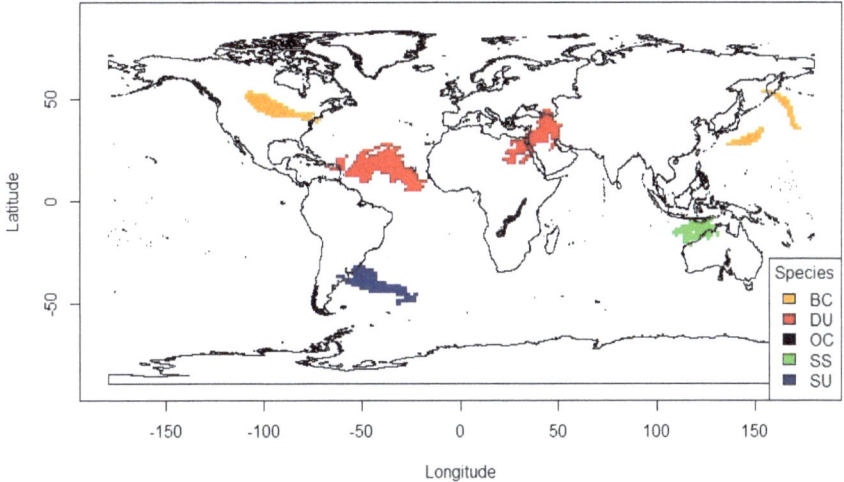

Fig. 2.3 Detected AARs availability of key aerosol species at an arbitrary time frame

in significant loss with 140,000 affected and 100,300 deaths, leading to respiratory illnesses and decreased visibility (Koplitz et al. 2016). The reconstruction of this event found there is presence of massive BC AAR which reduces the significant AQI of the region (Fig. 2.4a, b). The following year, Uttarakhand, India, faced Chir Pine forest fires, affecting 4,423 hectares of forest land and contributing to glacier melting and ecosystem disruption (Negi 2019). Similarly, this study found a BC AAR over the northern region of India (Fig. 2.4c, d). However, in 2018, British Columbia, experienced wildfires that impacted 1.35 million hectares of land, 56,000 cattle and calves displaced and increased the region's Air Quality Index (AQI) (Wang and Strong 2019). After this event, a BC AAR was found near the region shown in Fig. 2.4e. The 2020 wildfires in the southern region of Brazil devastated 5.07 billion USD, due to mortality, accounting for 0.68% of economic losses and equivalent to approximately 0.14% of Brazil's GDP (Wu et al. 2023) causing a public health crisis and ecological devastation. Similarly, this study found a BC AAR that covers the southern region of Brazil to Chille (Fig. 2.4g). However, some aerosols change the sky colour to orange. For example, in 2018, Northern India was struck by a dust storm, causing 125 deaths and leading to air pollution and reduced visibility (Sarkar et al. 2019). This study found a presence of DU AAR near the northern region of India during this event (Fig. 2.5a). Furthermore, a colossal dust storm called Godzilla dust storm from the Sahara Desert hit the regions of USA and Mexico in 2020, causing respiratory issues and impaired visibility (NASA; Chakraborty et al. 2021). Similarly, this study has detected DU AARs extending from Africa to USA during the event (Fig. 2.5c). In 2021, Also, a DU AAR was detected over East Asia which experienced a severe sandstorm, causing severe air pollution and visibility reduction (Fig. 2.5e) (Yu et al. 2023). Similarly, a DU AAR in 2022 was detected over Iraq when the region faced a dust storm coupled with La Niña, affecting 5,000 individuals and resulting in flight disruptions and health hazards (BBC 2022) (Fig. 2.5g). These events highlight the diverse impacts of natural disasters on regions worldwide, underscoring the importance of proactive measures to mitigate their effects and safeguard human lives and ecosystems.

2.5.5 Analysis of AAR-Induced Air Quality Deterioration and Associated Health Outcomes

The analysis of AAR-induced air quality deterioration and its associated health outcomes reveals the diverse impacts of aerosol pollution on human health in various regions worldwide. In regions such as southern Africa and eastern China, AAR-induced air pollution from sources like biomass burning and industrial activities has been linked to respiratory ailments, cardiovascular diseases, and premature mortality (Gordon et al. 2023; Amnuaylojaroen and Parasin 2023). The high concentration of aerosols carried by AARs exacerbates air quality issues, posing significant health risks to vulnerable populations, including children, the elderly, and individuals with pre-existing health conditions (Nagpure and Lal 2022). Similarly, in the northern hemisphere, particularly during winters, AAR-induced air pollution over the Indo-Gangetic-Brahmaputra (IGB) plains, eastern China, and Japan has been associated with adverse health outcomes (Pozzer et al. 2017; Ramachandran et al. 2020; Arima 2023). The combustion of fossil fuels, coupled with agricultural burning and industrial emissions, leads to elevated levels of particulate matter and other pollutants in the atmosphere, contributing to respiratory infections, cardiovascular problems, and premature death (Pandey et al. 2021). In regions with significant desertification, such as the northern side of the African continent, AAR-induced air pollution from dust storms poses serious health risks to populations living in affected areas (Mulcahy et al. 2014). The inhalation of fine dust particles carried by AARs can lead to respiratory issues, allergic reactions, and other health problems, particularly among vulnerable groups (Chatkin et al. 2022). The finding of this study underscores the urgent need for comprehensive measures to address AAR-induced air pollution and its detrimental effects on human health. Strategies such as reducing emissions from industrial sources, promoting clean energy technologies, and implementing air quality monitoring systems are essential for protecting public health and mitigating the impacts of aerosol pollution worldwide.

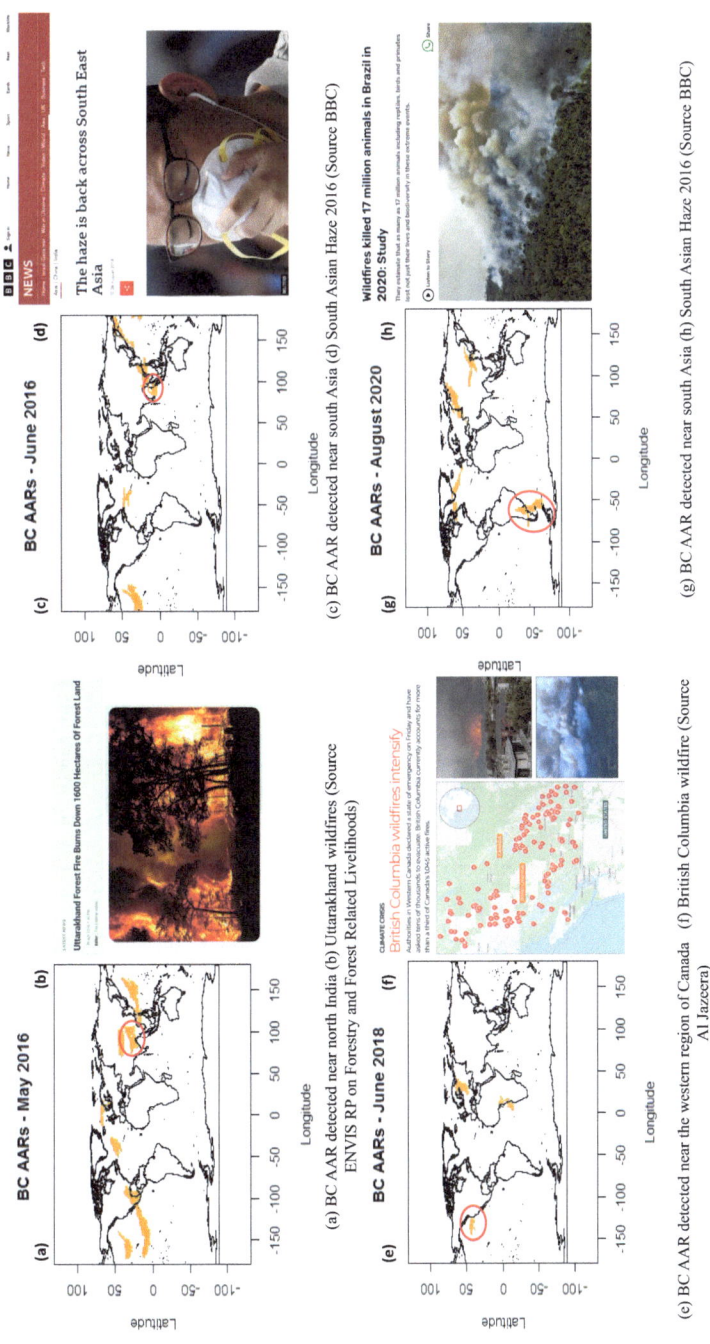

Fig. 2.4 Reconstruction of past extreme events and checking the availability of BC AARs

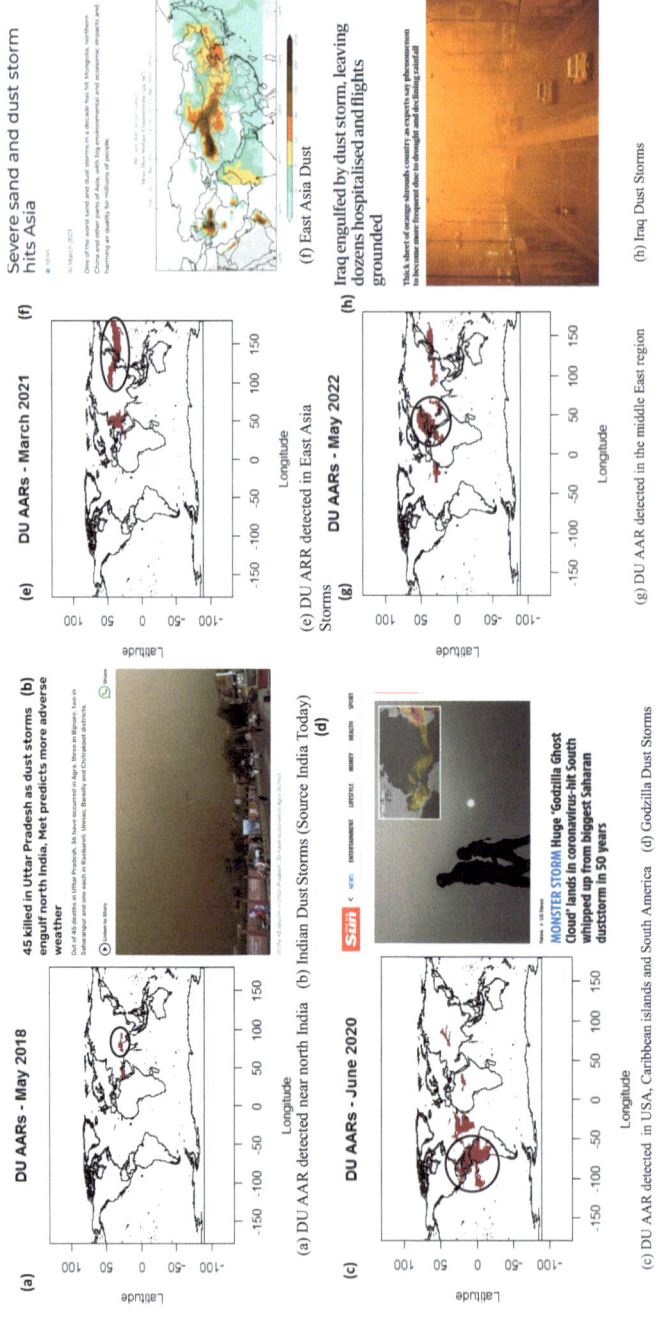

Fig. 2.5 Reconstruction of past extreme events and checking the availability of DU AARs

Table 2.2 Overview of major environmental events caused by AARs

Event	Year	Affected Regions	Cause	Loss	Impacts	References
1	2015	Equatorial Asia	Large-scale forest fires	140,000 affected, 100,300 deaths	Respiratory illness, decreased visibility	Koplitz et al. (2016)
2	2016	Uttarakhand, India	Chir Pine forest fires	4,423 hectares of forest land	Glacier melting, ecosystem disruption	Negi (2019)
3	2018	British Columbia	Wildfires	1.35 million hectares burned, 56,000 cattle and calves were located, increase AQI	Increased carbon emissions, global warming	Wang and Strong (2019)
4	2020	Brazil	Wildfires	5.07 billion USD, due to mortality, accounting for 0.68% of economic losses and equivalent to approximately 0.14% of Brazil's GDP	Public health crisis, ecological devastation	Wu et al. (2023)
5	2018	Northern India	Dust storm	125 deaths	Air pollution, reduced visibility	Sarkar et al. (2019)
6	2020	Sahara Desert, Africa	Colossal dust storm	-	Respiratory issues, impaired visibility	NASA (2020)
7	2021	East Asia	Sandstorm	-	Severe air pollution, visibility reduction	Yu et al. (2023)
8	2022	Iraq	Dust storms coupled with La Niña	5,000 affected, 1 death	Flight disruptions, health hazards	BBC (2022)

2.6 Mitigation and Adaptation Strategies

Addressing the complex challenges posed by AARs requires a multifaceted approach that encompasses policy interventions, public health measures, and community resilience-building initiatives (Watts et al. 2015; Randall et al. 2021). Policy interventions aimed at reducing aerosol emissions play a crucial role in mitigating AAR-related health risks (Amann et al. 2020). Governments and regulatory bodies must implement stringent air quality standards and enforce emission control measures for industries, vehicles, and other sources of pollution (Purohit et al. 2019; Amann et al. 2020). Public health initiatives have a critical role in reducing exposure to AARs and lessening their negative health effects in concert with governmental efforts (Giles et al. 2011). Additionally, international cooperation and agreements are essential for addressing transboundary aerosol pollution and mitigating its adverse effects on public health (Yamineva and Romppanen 2017). Public health initiatives have a critical role in reducing exposure to AARs and lessening their negative health effects in concert with governmental efforts. Public health initiatives have a critical role in reducing exposure to AARs and lessening their negative health effects in concert with governmental efforts. Public awareness campaigns can educate communities about the risks associated with aerosol pollution and provide guidance on preventive measures, such as staying indoors during periods of poor air quality and using air purifiers to reduce indoor pollution levels. Healthcare infrastructure must be strengthened to cope with the increased burden of respiratory illnesses and cardiovascular diseases resulting from AAR-induced air pollution (Lee et al. 2014). This includes ensuring access to medical care, respiratory protective equipment, and vaccinations to mitigate the spread of respiratory infections during haze episodes and dust storms.

Furthermore, community resilience-building initiatives are essential, particularly in regions prone to AAR-related health hazards (Chandra et al. 2011). Local authorities and community organizations should develop emergency response plans and early warning systems to alert residents about impending environmental crises, such as dust storms and haze events (Rogers and Tsirkunov). Community-based initiatives, such as tree planting campaigns and urban greening projects, can help mitigate the impacts of aerosol pollution by enhancing air quality and reducing the urban heat island effect. Moreover, efforts to diversify livelihoods and promote sustainable agriculture practices can reduce dependence on activities that contribute to aerosol emissions, such as biomass burning and land clearing.

2.7 Future Directions and Research Needs

Since AAR is a newly developed concept and research AARs continue to evolve, several emerging trends have profound implications for human health. Based on the findings, one notable area of focus is the integration of advanced technologies, such as Artificial Intelligence (AI) and Machine Learning (ML), into AAR research to enhance forecasting capabilities (Singh and Goyal 2023a; Shakya et al. 2023; Meghani et al. 2023; Rautela et al. 2024b). By leveraging AI and ML algorithms, researchers can analyze vast datasets of meteorological and aerosol variables to improve the accuracy and precision of AAR predictions (Shakya et al. 2023). These AI-driven forecasting models have the potential to provide early warnings of AAR events, enabling policymakers and public health officials to implement timely interventions to mitigate the impacts of aerosol pollution on human health. Furthermore, there is a pressing need for interdisciplinary collaboration to address the complex challenges posed by AAR-related health impacts. Collaborative efforts between meteorologists, atmospheric scientists, public health experts, and policy-makers can facilitate a comprehensive understanding of the mechanisms driving AAR-induced air pollution and its effects on human health. By combining exper-tise from diverse disciplines, researchers can develop holistic approaches to assess and mitigate the health risks associated with AARs, ultimately safeguarding the well-being of populations exposed to aerosol pollution.

Areas requiring further investigation and advancements in understanding AAR-related health impacts include the characterization of aerosol composition, sources, and transport mechanisms. Comprehensive studies are needed to elucidate the chem-ical composition of aerosols carried by AARs and their interactions with atmo-spheric conditions. Additionally, advancements in remote sensing technologies and observational techniques can provide valuable insights into the spatial and temporal distribution of aerosol pollution, enabling more accurate assessments of its health impacts. Moreover, future research should explore the long-term health effects of AAR-induced air pollution, particularly among vulnerable populations. Epidemio-logical studies examining the relationship between chronic exposure to aerosol pollu-tion and respiratory diseases, cardiovascular disorders, and other health outcomes can inform targeted interventions and public health policies. Additionally, efforts to develop early warning systems and adaptive strategies to mitigate the health risks associated with AARs are essential for enhancing resilience to environmental hazards in the face of climate change.

2.8 Conclusion

The study of Aerosol Atmospheric Rivers (AARs) represents a critical frontier in environmental research, with profound implications for human health and well-being. This comprehensive analysis has delved into various aspects of AARs, beginning with an exploration of their formation and dynamics. AARs, akin to their atmospheric counterparts, serve as conduits for the long-distance transport of aerosols across the atmosphere. These aerosols, originating from both natural phenomena and anthro-pogenic activities, contribute to air quality degradation and pose significant health risks. Understanding the types and sources of aerosols transported by AARs is essen-tial for assessing their environmental and health impacts. From black carbon and dust to organic carbon and sea salt, each aerosol type presents unique inhalation hazards and health effects. Moreover, the distribution and intensity of these aerosols vary by season and region, driven by complex interactions between natural processes, human activities, and atmospheric dynamics. The health impacts of AAR-induced air pollution are profound and wide-ranging, affecting vulnerable populations and regions disproportionately. Short-term exposure to high concentrations of aerosols can lead to acute respiratory symptoms and exacerbate existing health conditions, while chronic exposure is associated with long-term respiratory diseases, cardio-vascular disorders, and other adverse health outcomes. Mitigation and adaptation strategies are imperative to address these challenges, encompassing policy interventions, public health measures, and community resilience-building initiatives. Moving forward, future research directions and needs in AAR studies are outlined to enhance our understanding and response to this environmental phenomenon. Integration of advanced technologies such as Artificial Intelligence (AI) and Machine Learning (ML) into AAR research holds promise for improving forecasting capabilities and informing timely interventions. Interdisciplinary collaboration among meteorologists, atmospheric scientists, public health experts, and policymakers is crucial for developing holistic approaches to assess and mitigate the health risks associated with AARs. Characterization of aerosol composition, sources, and transport mechanisms, along with studies on long-term health effects and vulnerable populations, are identified as priority areas for further investigation. Additionally, the development of early warning systems and adaptive strategies to mitigate AAR-induced air pollution is essential for enhancing resilience to environmental hazards in the face of climate change.

References

Amann M, Kiesewetter G, Schöpp W et al (2020) Reducing global air pollution: the scope for further policy interventions. Philos Trans R Soc A Math Phys Eng Sci 378:20190331. https://doi.org/10.1098/rsta.2019.0331

Amnuaylojaroen T, Parasin N (2023) Perspective on particulate matter: from biomass burning to the health crisis in Mainland Southeast Asia. Toxics 11:553. https://doi.org/10.3390/toxics11070553

Arima H (2023) Seasonal variation in air pollutant levels and its effects on the sex ratio at birth on Fukue island. Japan. BMC Public Health 23:2471. https://doi.org/10.1186/s12889-023-17418-5

Artaxo P, Hansson H-C, Andreae MO, et al (2022) Tropical and Boreal forest—atmosphere interactions: a review. Tellus B Chem Phys Meteorol 74:24. https://doi.org/10.16993/tellusb.34

Basith S, Manavalan B, Shin TH, et al (2022) The impact of fine particulate matter 2.5 on the cardiovascular system: a review of the invisible killer. Nanomaterials 12:2656. https://doi.org/10.3390/nano12152656

BBC (2022) Iraq dust storm leaves 5,000 people needing treatment—BBC News. https://www.bbc.com/news/world-middle-east-61335124. Accessed 4 Jul 2023

Biddle TA, Li Q, Maltz MR et al (2021) Salton Sea aerosol exposure in mice induces a pulmonary response distinct from allergic inflammation. Sci Total Environ 792:148450. https://doi.org/10.1016/j.scitotenv.2021.148450

Bozkurt D, Sen OL, Ezber Y, et al (2021) Influence of African atmospheric rivers on precipitation and snowmelt in the near East's Highlands. J Geophys Res Atmos 126.https://doi.org/10.1029/2020JD033646

Chakraborty S, Guan B, Waliser DE, et al (2021) Extending the atmospheric river concept to aerosols: climate and air quality impacts. Geophys Res Lett 48.https://doi.org/10.1029/2020GL091827

Chakraborty S, Guan B, Waliser DE, da Silva AM (2022) Aerosol atmospheric rivers: climatology, event characteristics, and detection algorithm sensitivities. Atmos Chem Phys 22:8175–8195. https://doi.org/10.5194/acp-22-8175-2022

Chakraborty S, Schiro KA, Fu R, Neelin JD (2018) On the role of aerosols, humidity, and vertical wind shear in the transition of shallow-to-deep convection at the Green Ocean Amazon 2014/5 site. Atmos Chem Phys 18:11135–11148. https://doi.org/10.5194/acp-18-11135-2018

Chandra A, Acosta J, Howard S et al (2011) Building community resilience to disasters: a way forward to enhance national health security. Rand Heal Q 1:6

Chatkin J, Correa L, Santos U (2022) External Environmental pollution as a risk factor for Asthma. Clin Rev Allergy Immunol 62:72–89. https://doi.org/10.1007/s12016-020-08830-5

Che H, Xia X, Zhao H et al (2024) Aerosol optical and radiative properties and their environmental effects in China: a review. Earth-Sci Rev 248:104634. https://doi.org/10.1016/j.earscirev.2023.104634

Chen J, Li C, Ristovski Z et al (2017) A review of biomass burning: emissions and impacts on air quality, health and climate in China. Sci Total Environ 579:1000–1034. https://doi.org/10.1016/j.scitotenv.2016.11.025

Christensen MW, Gettelman A, Cermak J et al (2022) Opportunistic experiments to constrain aerosol effective radiative forcing. Atmos Chem Phys 22:641–674. https://doi.org/10.5194/acp-22-641-2022

Dai Q, Bi X, Song W et al (2019) Residential coal combustion as a source of primary sulfate in Xi'an, China. Atmos Environ 196:66–76. https://doi.org/10.1016/j.atmosenv.2018.10.002

Dubey S, Goyal MK (2020) Glacial lake outburst flood hazard, downstream impact, and risk over the Indian Himalayas. Water Resour Res 56.https://doi.org/10.1029/2019WR026533

Filkov AI, Ngo T, Matthews S et al (2020) Impact of Australia's catastrophic 2019/20 bushfire season on communities and environment. Retrospective analysis and current trends. J Saf Sci Resil 1:44–56. https://doi.org/10.1016/j.jnlssr.2020.06.009

Filonchyk M, Yan H, Zhang Z et al (2019) Combined use of satellite and surface observations to study aerosol optical depth in different regions of China. Sci Rep 9:6174. https://doi.org/10.1038/s41598-019-42466-6

Fussell JC, Kelly FJ (2021) Mechanisms underlying the health effects of desert sand dust. Environ Int 157:106790. https://doi.org/10.1016/j.envint.2021.106790

Gelaro R, McCarty W, Suárez MJ et al (2017) The modern-era retrospective analysis for research and applications, version 2 (MERRA-2). J Clim 30:5419–5454. https://doi.org/10.1175/JCLI-D-16-0758.1

Giles LV, Barn P, Künzli N et al (2011) From good intentions to proven interventions: effectiveness of actions to reduce the health impacts of air pollution. Environ Health Perspect 119:29–36. https://doi.org/10.1289/ehp.1002246

Gollakota ARK, Gautam S, Santosh M et al (2021) Bioaerosols: characterization, pathways, sampling strategies, and challenges to geo-environment and health. Gondwana Res 99:178–203. https://doi.org/10.1016/j.gr.2021.07.003

Gordon JND, Bilsback KR, Fiddler MN, et al (2023) The effects of trash, residential biofuel, and open biomass burning emissions on local and transported PM 2.5 and Its attributed mortality in Africa. GeoHealth 7. https://doi.org/10.1029/2022GH000673

Gupta T, Rajeev P, Rajput R (2022) Emerging major role of organic aerosols in explaining the occurrence, frequency, and magnitude of haze and Fog episodes during wintertime in the indo Gangetic plain. ACS Omega 7:1575–1584. https://doi.org/10.1021/acsomega.1c05467

Gurjar BR, Ravindra K, Nagpure AS (2016) Air pollution trends over Indian megacities and their local-to-global implications. Atmos Environ 142:475–495. https://doi.org/10.1016/j.atmosenv.2016.06.030

Hirdman D, Burkhart JF, Sodemann H et al (2010) Long-term trends of black carbon and sulphate aerosol in the Arctic: changes in atmospheric transport and source region emissions. Atmos Chem Phys 10:9351–9368. https://doi.org/10.5194/acp-10-9351-2010

Horowitz HM, Holmes C, Wright A, et al (2020) Effects of sea salt aerosol emissions for marine cloud brightening on atmospheric chemistry: implications for radiative forcing. Geophys Res Lett 47.https://doi.org/10.1029/2019GL085838

Hu Y, Kang S, Yang J et al (2022) Impact of atmospheric circulation patterns on properties and regional transport pathways of aerosols over Central-West Asia: emphasizing the Tibetan Plateau. Atmos Res 266:105975. https://doi.org/10.1016/j.atmosres.2021.105975

Jacob DJ, Waldman JM, Munger JW, Hoffmann MR (1986) The H_2SO_4–HNO_3–NH_3 system at high humidities and in fogs. 2. Comparison of field data with thermodynamic calculations. J Geophys Res 91:1089–1096. https://doi.org/10.1029/JD091iD01p01089

Jha S, Das J, Goyal MK (2019) Assessment of risk and resilience of terrestrial ecosystem productivity under the influence of extreme climatic conditions over India. Sci Rep 9:18923. https://doi.org/10.1038/s41598-019-55067-0

Jiang XQ, Mei XD, Feng D (2016) Air pollution and chronic airway diseases: what should people know and do? J Thorac Dis 8:E31–E40. https://doi.org/10.3978/J.ISSN.2072-1439.2015.11.50

Karydis VA, Kumar P, Barahona D, et al (2011) On the effect of dust particles on global cloud condensation nuclei and cloud droplet number. J Geophys Res Atmos 116:n/a-n/a. https://doi.org/10.1029/2011JD016283

Kashyap P, Kumar A, Kumar RP, Kumar K (2019) Biogenic and anthropogenic isoprene emissions in the subtropical urban atmosphere of Delhi. Atmos Pollut Res 10:1691–1698. https://doi.org/10.1016/j.apr.2019.07.004

Koplitz SN, Mickley LJ, Marlier ME et al (2016) Public health impacts of the severe haze in Equatorial Asia in September–October 2015: demonstration of a new framework for informing fire management strategies to reduce downwind smoke exposure. Environ Res Lett 11:094023. https://doi.org/10.1088/1748-9326/11/9/094023

Kumar M, Kumari A, Kushwaha DP et al (2020) Estimation of daily stage-discharge relationship by using data-driven techniques of a perennial River. India. Sustainability 12:7877. https://doi.org/10.3390/su12197877

Lapere R, Thomas JL, Favier V, et al (2024) Polar aerosol atmospheric rivers: detection, characteristics, and potential applications. J Geophys Res Atmos 129.https://doi.org/10.1029/2023JD039606

Lau WKM, Kim K-M, Shi J-J et al (2017) Impacts of aerosol–monsoon interaction on rainfall and circulation over Northern India and the Himalaya Foothills. Clim Dyn 49:1945–1960. https://doi.org/10.1007/s00382-016-3430-y

Lee B-J, Kim B, Lee K (2014) Air pollution exposure and cardiovascular disease. Toxicol Res 30:71–75. https://doi.org/10.5487/TR.2014.30.2.071

Li J, Carlson BE, Yung YL et al (2022) Scattering and absorbing aerosols in the climate system. Nat Rev Earth Environ 3:363–379. https://doi.org/10.1038/s43017-022-00296-7

Loucks DP, van Beek E (2017) Water resources planning and management: an overview. Water resource systems planning and management. Springer International Publishing, Cham, pp 1–49

Manisalidis I, Stavropoulou E, Stavropoulos A, Bezirtzoglou E (2020) Environmental and Health Impacts of Air Pollution: A Review. Front Public Heal 8.https://doi.org/10.3389/fpubh.2020.00014

May NW, Bernays N, Farley R et al (2023) Intensive aerosol properties of boreal and regional biomass burning aerosol at Mt. Bachelor Observatory: larger and black carbon (BC)-dominant particles transported from Siberian wildfires. Atmos Chem Phys 23:2747–2764. https://doi.org/10.5194/acp-23-2747-2023

Meghani S, Singh S, Kumar N, Goyal MK (2023) Predicting the spatiotemporal characteristics of atmospheric rivers: a novel data-driven approach. Glob Planet Change 231:104295. https://doi.org/10.1016/j.gloplacha.2023.104295

Merdji AB, Lu C, Xu X, Mhawish A (2023) Long-term three-dimensional distribution and transport of Saharan dust: observation from CALIPSO, MODIS, and reanalysis data. Atmos Res 286:106658. https://doi.org/10.1016/j.atmosres.2023.106658

Mulcahy JP, Walters DN, Bellouin N, Milton SF (2014) Impacts of increasing the aerosol complexity in the Met Office global numerical weather prediction model. Atmos Chem Phys 14:4749–4778. https://doi.org/10.5194/acp-14-4749-2014

Nagpure AS, Lal RM (2022) PM 2.5 exposures increased for the majority of Indians and a third of the global population during COVID-19 lockdowns: a residential biomass burning and environmental justice perspective. Environ Res Lett 17:114017. https://doi.org/10.1088/1748-9326/ac9703

Nair SK, Parameswaran K, Rajeev K (2005) Seven year satellite observations of the mean structures and variabilities in the regional aerosol distribution over the oceanic areas around the Indian subcontinent. Ann Geophys 23:2011–2030. https://doi.org/10.5194/angeo-23-2011-2005

Negi GCS (2019) Forest fire in Uttarakhand: causes, consequences and remedial measures. Int J Ecol Environ Sci 45:31–37

Oh H-J, Ma Y, Kim J (2020) Human inhalation exposure to aerosol and health effect: aerosol monitoring and modelling regional deposited doses. Int J Environ Res Public Health 17:1923. https://doi.org/10.3390/ijerph17061923

Ovadnevaite J, Manders A, de Leeuw G et al (2014) A sea spray aerosol flux parameterization encapsulating wave state. Atmos Chem Phys 14:1837–1852. https://doi.org/10.5194/acp-14-1837-2014

Palacio LC, Pachajoa DC, Echeverri-Londoño CA, et al (2023) Air pollution and cardiac diseases: a review of experimental studies. Dose-Response 21.https://doi.org/10.1177/15593258231212793

Pandey A, Brauer M, Cropper ML et al (2021) Health and economic impact of air pollution in the states of India: the global burden of disease study 2019. Lancet Planet Heal 5:e25–e38. https://doi.org/10.1016/S2542-5196(20)30298-9

Persad GG (2023) The dependence of aerosols' global and local precipitation impacts on the emitting region. Atmos Chem Phys 23:3435–3452. https://doi.org/10.5194/acp-23-3435-2023

Poonia V, Goyal MK, Gupta BB et al (2021) Drought occurrence in different river basins of India and blockchain technology based framework for disaster management. J Clean Prod 312:127737. https://doi.org/10.1016/j.jclepro.2021.127737

Pope CA, Burnett RT, Thurston GD et al (2004) Cardiovascular mortality and long-term exposure to particulate air pollution. Circulation 109:71–77. https://doi.org/10.1161/01.CIR.0000108927.80044.7F

Pöschl U (2005) Atmospheric aerosols: composition, transformation, climate and health effects. Angew Chemie Int Ed 44:7520–7540. https://doi.org/10.1002/anie.200501122

Pozzer A, Tsimpidi AP, Karydis VA et al (2017) Impact of agricultural emission reductions on fine-particulate matter and public health. Atmos Chem Phys 17:12813–12826. https://doi.org/10.5194/acp-17-12813-2017

Purohit P, Amann M, Kiesewetter G et al (2019) Mitigation pathways towards national ambient air quality standards in India. Environ Int 133:105147. https://doi.org/10.1016/j.envint.2019.105147

Qian Y, Flanner MG, Leung LR, Wang W (2011) Sensitivity studies on the impacts of Tibetan Plateau snowpack pollution on the Asian hydrological cycle and monsoon climate. Atmos Chem Phys 11:1929–1948. https://doi.org/10.5194/acp-11-1929-2011

Ralph FM, Waliser DE, Dettinger MD et al (2020) The future of atmospheric river research and applications. Atmospheric Rivers. Springer International Publishing, Cham, pp 219–247

Ramachandran S, Rupakheti M, Lawrence MG (2020) Black carbon dominates the aerosol absorption over the Indo-Gangetic Plain and the Himalayan foothills. Environ Int 142:105814. https://doi.org/10.1016/j.envint.2020.105814

Ramanathan V, Crutzen PJ, Kiehl JT, Rosenfeld D (2001) Aerosols, climate, and the hydrological cycle. Science (80) 294:2119–2124. https://doi.org/10.1126/science.1064034

Randall K, Ewing ET, Marr LC, et al (2021) How did we get here: what are droplets and aerosols and how far do they go? A historical perspective on the transmission of respiratory infectious diseases. Interface Focus 11.https://doi.org/10.1098/rsfs.2021.0049

Randles CA, da Silva AM, Buchard V et al (2017) The MERRA-2 aerosol Reanalysis, 1980 onward. Part I: system description and data assimilation evaluation. J Clim 30:6823–6850. https://doi.org/10.1175/JCLI-D-16-0609.1

Rasch PJ, Tilmes S, Turco RP et al (2008) An overview of geoengineering of climate using strato-spheric sulphate aerosols. Philos Trans R Soc A Math Phys Eng Sci 366:4007–4037. https://doi.org/10.1098/rsta.2008.0131

Rautela KS, Kumar D, Gandhi BGR et al (2023a) Long-term hydrological simulation for the estimation of snowmelt contribution of Alaknanda River Basin, Uttarakhand using SWAT. J Water Supply Res Technol 72:139–159. https://doi.org/10.2166/aqua.2023.176

Rautela KS, Kumar D, Gandhi BGR et al (2023b) Evaluating hydroelectric potential in Alaknanda basin, Uttarakhand using the snowmelt runoff model (SRM). J Water Clim Chang 14:4146–4161. https://doi.org/10.2166/wcc.2023.341

Rautela KS, Kuniyal JC, Goyal MK et al (2024a) Assessment and modelling of hydro-sedimentological flows of the eastern river Dhauliganga, north-western Himalaya. Nat Hazards, India. https://doi.org/10.1007/s11069-024-06413-7

Rautela KS, Singh S, Goyal MK (2024b) Characterizing the spatio-temporal distribution, detection, and prediction of aerosol atmospheric rivers on a global scale. J Environ Manage 351:119675. https://doi.org/10.1016/j.jenvman.2023.119675

Rautela KS, Singh S, Goyal MK (2024c) Resilience to air pollution: A novel approach for detecting and predicting aerosol atmospheric rivers within earth system boundaries earth systems and environment. https://doi.org/10.1007/s41748-024-00421-0

Rentschler J, Leonova N (2023) Global air pollution exposure and poverty. Nat Commun 14:4432. https://doi.org/10.1038/s41467-023-39797-4

Rogers D, Tsirkunov V, This report summarizes how to implement multi-hazard early warning systems based on best practices in the hydrometeorological community Implementing Hazard Early Warning Systems GFDRR WCIDS Report 11–03

Rosenfeld D, Zheng Y, Hashimshoni E et al (2016) Satellite retrieval of cloud condensation nuclei concentrations by using clouds as CCN chambers. Proc Natl Acad Sci 113:5828–5834. https://doi.org/10.1073/pnas.1514044113

Salin J, Ohtonen P, Andersson MA, Syrjälä H (2021) The toxicity of wiped dust and airborne microbes in individual classrooms increase the risk of teachers' work-related symptoms: a cross-sectional study. Pathogens 10:1360. https://doi.org/10.3390/pathogens10111360

Sarkar S, Chauhan A, Kumar R, Singh RP (2019) Impact of deadly dust storms (May 2018) on air quality, meteorological, and atmospheric parameters over the northern parts of India. GeoHealth 3:67–80. https://doi.org/10.1029/2018GH000170

Shakya D, Deshpande V, Goyal MK, Agarwal M (2023) PM2.5 air pollution prediction through deep learning using meteorological, vehicular, and emission data: a case study of New Delhi, India. J Clean Prod 427:139278. https://doi.org/10.1016/j.jclepro.2023.139278

Singh S, Goyal MK (2023a) An innovative approach to predict atmospheric rivers: exploring convolutional autoencoder. Atmos Res 289:106754. https://doi.org/10.1016/j.atmosres.2023.106754

Singh S, Goyal MK (2023b) Enhancing climate resilience in businesses: the role of artificial intelligence. J Clean Prod 418:138228. https://doi.org/10.1016/j.jclepro.2023.138228

Singh S, Goyal MK, Jha S (2023a) Role of large-scale climate oscillations in precipitation extremes associated with atmospheric rivers: nonstationary framework. Hydrol Sci J 68:395–411. https://doi.org/10.1080/02626667.2022.2159412

Singh S, Goyal MK, Jha S Role of large-scale climate oscillations in precipitation extremes associated with atmospheric rivers: nonstationary framework. Hydrol Sci J. https://doi.org/10.1080/02626667.2022.2159412

Singh S, Kumar N, Goyal MK, Jha S (2023b) Relative influence of ENSO, IOD, and AMO over spatiotemporal variability of hydroclimatic extremes in Narmada basin, India. AQUA—water infrastructure. Ecosyst Soc 72:520–539. https://doi.org/10.2166/aqua.2023.219

Sinha J, Jha S, Goyal MK (2019) Influences of watershed characteristics on long-term annual and intra-annual water balances over India. J Hydrol 577:123970. https://doi.org/10.1016/j.jhydrol.2019.123970

Sofi MS, Rautela KS, Bhat SU et al (2021) Application of geomorphometric approach for the estimation of hydro-sedimentological flows and cation weathering rate: towards understanding the sustainable land use policy for the Sindh Basin, Kashmir Himalaya. Water, Air, Soil Pollut 232:280. https://doi.org/10.1007/s11270-021-05217-w

Tian Y, Duan M, Cui X, et al (2023) Advancing application of satellite remote sensing technologies for linking atmospheric and built environment to health. Front Public Heal 11.https://doi.org/10.3389/fpubh.2023.1270033

Wang J, Strong K (2019) British Columbia's forest fires, 2018. Stat Canada 8

Watts N, Adger WN, Agnolucci P et al (2015) Health and climate change: policy responses to protect public health. Lancet 386:1861–1914. https://doi.org/10.1016/S0140-6736(15)60854-6

Wen W, Ma X, Guo C et al (2020) The aerosol-radiation interaction effects of different particulate matter components during heavy pollution periods in China. Atmosphere (basel) 11:254. https://doi.org/10.3390/atmos11030254

Wu T, Boor BE (2021) Urban aerosol size distributions: a global perspective. Atmos Chem Phys 21:8883–8914. https://doi.org/10.5194/acp-21-8883-2021

Wu Y, Li S, Xu R, et al (2023) Wildfire-related PM2.5 and health economic loss of mortality in Brazil. Environ Int 174:107906. https://doi.org/10.1016/j.envint.2023.107906

Xu X, Wu H, Yang X, Xie L (2020) Distribution and transport characteristics of dust aerosol over Tibetan Plateau and Taklimakan Desert in China using MERRA-2 and CALIPSO data. Atmos Environ 237:117670. https://doi.org/10.1016/j.atmosenv.2020.117670

Yamineva Y, Romppanen S (2017) Is law failing to address air pollution? Reflections on international and <scp>EU</scp> developments. Rev Eur Comp Int Environ Law 26:189–200. https://doi.org/10.1111/reel.12223

Yan D, Lei Y, Shi Y et al (2018) Evolution of the spatiotemporal pattern of PM2.5 concentrations in China—a case study from the Beijing-Tianjin-Hebei region. Atmos Environ 183:225–233. https://doi.org/10.1016/j.atmosenv.2018.03.041

Yang J, Kang S, Ji Z, Chen D (2018) Modeling the origin of anthropogenic black carbon and its climatic effect over the Tibetan plateau and surrounding regions. J Geophys Res Atmos 123:671–692. https://doi.org/10.1002/2017JD027282

Yu T, Xiaole P, Yujie J et al (2023) East Asia dust storms in spring 2021: transport mechanisms and impacts on China. Atmos Res 290:106773. https://doi.org/10.1016/j.atmosres.2023.106773

Zhang B (2020) The effect of aerosols to climate change and society. J Geosci Environ Prot 08:55–78. https://doi.org/10.4236/gep.2020.88006

Zhang X, Wang H, Che H-Z et al (2021) Radiative forcing of the aerosol-cloud interaction in seriously polluted East China and East China Sea. Atmos Res 252:105405. https://doi.org/10.1016/j.atmosres.2020.105405

Zhao H, Gui K, Ma Y, et al (2022) Effects of different aerosols on the air pollution and their relationship with meteorological parameters in North China Plain. Front Environ Sci 10.https://doi.org/10.3389/fenvs.2022.814736

Chapter 3
Application of Data Mining and AI&ML in Aerosol Pollution and Aerosol Atmospheric Rivers

Abstract In recent years, the application of data mining, artificial intelligence, and machine learning (AI&ML) techniques has revolutionized the field of earth and environmental science, particularly in the study of atmospheric pollutants such as aerosol pollution and aerosol atmospheric rivers (AARs). This chapter provides a comprehensive overview of the diverse applications of data mining and AI&ML techniques in aerosol science, highlighting recent advancements, methodological approaches, and a case study to predict the spatio-temporal patterns of AARs using convolutional autoencoders. Data mining techniques enable the extraction of valuable insights and patterns from large environmental datasets, while AI&ML techniques facilitate the development of predictive models for aerosol concentrations, dispersion patterns, and atmospheric interactions. Clustering and classification algorithms identify aerosol pollution hotspots and predict pollution events, while association rule mining techniques reveal correlations between aerosol pollution and meteorological variables. AI&ML models, including neural networks and support vector machines, forecast aerosol concentration levels and classify pollution severity, aiding in air quality forecasting and early warning systems. The case study demonstrates the efficacy of these techniques in identifying aerosol pollution patterns, predicting AARs, and integrating satellite data for global aerosol monitoring. Despite challenges such as data quality and model interpretability, future research directions aim to enhance data accessibility, improve integration techniques, and explore emerging AI&ML methodologies for more accurate predictions and informed decision-making in aerosol pollution management.

Keywords Aerosol pollution · AI&ML · AARs · Data mining · Predictive modelling

M. K. Goyal and K. S. Rautela, *Aerosol Atmospheric Rivers*,
SpringerBriefs in Applied Sciences and Technology,
https://doi.org/10.1007/978-3-031-66758-9_3

3.1 Introduction

In recent years, the field of earth and environmental science has witnessed a surge in interest and innovation surrounding the application of data mining, artificial intelligence, and machine learning (AI&ML) techniques (Goyal 2014; Sekar et al. 2016; Gibert et al. 2018; Krishan et al. 2019; Singh and Goyal 2023a; Meghani et al. 2023). These advanced analytical tools hold significant promise for addressing complex environmental challenges, including the study of atmospheric pollutants (Ma et al. 2021), aerosol pollution (Li et al. 2022a), and aerosol atmospheric rivers (Rautela et al. 2024a). However, in the context of aerosol science, the application of data mining and AI&ML techniques offers unprecedented opportunities to enhance our understanding of aerosol dynamics, pollution sources, and their impacts on environmental and human health (Haque and Singh 2017; Mo and Li 2019; Oh et al. 2020; Liu et al. 2020).

Data mining is the process of using several computing techniques to extract useful knowledge, patterns, and insights from massive datasets (Ojha et al. 2007; Goyal and Ojha 2011; Goyal et al. 2012; Senthil Kumar et al. 2013; Kotu and Deshpande 2015; Goyal and Sharma 2016). With the exponential growth of data collection and storage capabilities, data mining techniques have become indispensable tools for analysing complex environmental datasets, including those related to aerosol pollution (Bellinger et al. 2017). Through the utilization of sophisticated algorithms like clustering, classification, and association rule mining, researchers are able to detect both temporal and spatial patterns in aerosol concentrations, as well as underlying trends and factors that contribute to aerosol pollution (Bin Tarek et al. 2018; Zhang et al. 2020; Hu et al. 2021). Conversely, AI&ML include a collection of statistical methods and computational algorithms that let computers learn from data, anticipate future events, and adjust to changing surroundings without the need for explicit programming (Elahi et al. 2023). AI&ML approaches have special benefits for modelling intricate aerosol processes, forecasting future trends, and evaluating the efficacy of mitigation measures in the context of aerosol science (Mo and Li 2019; Qiu et al. 2022, 2023; Peng et al. 2024). Accurate predictive models that are trained based on observational data, satellite imaging, and atmospheric simulations will be prepared to forecast aerosol concentrations, dispersion patterns, and atmospheric interactions (Peng et al. 2024).

A wide range of research areas are involved in the application of data mining and AI&ML techniques in aerosol science, all of which further our knowledge of aerosol dynamics and their effects on human and environmental health. In order to differentiate between human and natural emissions, aerosol source apportionment uses data mining techniques like factor analysis and receptor modelling to locate and measure the sources of aerosol pollution (Kulmala et al. 2023). AI&ML algorithms, including neural networks, SVM, and random forests, are utilized for aerosol prediction and forecasting, aiding in air quality forecasting and early warning systems (Qiu et al. 2023). Recent progressions in AI&ML models have become indispensable for the prediction of earth and atmospheric variables (Qiu et al. 2023). These models offer predictive capabilities at both regional and global scales for aerosol transport (Qiu et al. 2023), particulate matter (PM) (Shakya et al. 2023), and assessing patterns of atmospheric rivers (ARs) (Singh and Goyal 2023a, b). These models are highly effective at simulating intricate interactions that span various spatio-temporal dimensions and include both linear and nonlinear dynamics (Yu et al. 2019; Alzubaidi et al. 2021).

Highlighting new developments, methodological strategies, and case studies from the literature, this chapter will examine the various uses of data mining and AI&ML techniques in aerosol science. These results will enhance air quality management plans, learn more about aerosol dynamics, and eventually protect the environment and public health from the damaging effects of aerosol pollution by utilizing data-driven approaches. However, a paradigm shift in this approach to the study of aerosol pollution through aerosol atmospheric rivers is provided by the combination of data mining and AI&ML approaches.

3.2 Data Mining Techniques for Aerosol Pollution Analysis

3.2.1 Data Sources for Aerosol Pollution Data

Aerosol pollution data are derived from various sources, including ground-based monitoring stations, satellite observations, and atmospheric modelling simulations (Filonchyk et al. 2019; Li et al. 2022b). Ground-based monitoring stations provide direct measurements of aerosol concentrations, typically collected using instruments such as nephelometers and particle counters (Kuniyal and Guleria 2019). These data offer valuable insights into local aerosol pollution levels and spatial variability. Satellite observations, on the other hand, offer a broader perspective, capturing aerosol distributions over large geographic areas (Christensen et al. 2020; Masoud 2023). Sensors onboard satellites, such as Multiangle imaging spectroradiometer (MISR), Moderate Resolution Imaging Spectroradiometer (MODIS), and Cloud-Aerosol Lidar and Infrared Pathfinder Satellite Observation (CALIPSO), provide measurements of aerosol optical properties, allowing for the estimation of aerosol concentrations and identification of aerosol sources (Brun et al. 2011; Sahak et al. 2019). Atmospheric modelling simulations, including chemical transport models (CTMs) and numerical weather prediction (NWP) models coupled with AI&ML models, simulate aerosol transport and transformation processes with the integration of meteorological variables with emission inventories to generate spatiotemporally resolved aerosol concentration fields (Prank et al. 2016; Rojas et al. 2020).

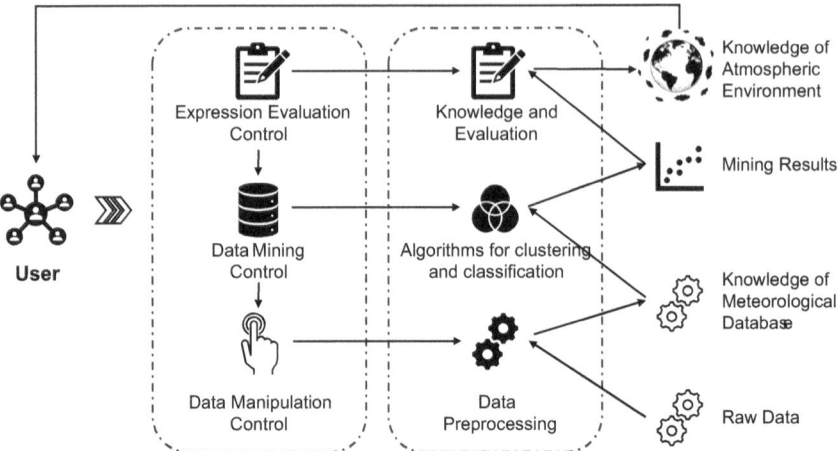

Fig. 3.1 Process of data mining for atmospheric environment

3.2.2 Overview of Data Mining Techniques

A wide number of approaches are included in data mining techniques, which are used to uncover links, patterns, and insights from big, complicated databases (Fig. 3.1) (Bellinger et al. 2017). These methods fall into two general categories: supervised and unsupervised learning approaches (Chowdhury et al. 2020; Nandakumar 2022; Essamlali et al. 2024). Aerosol pollution hotspots can be found using unsupervised learning approaches like clustering algorithms, which find natural groupings or clusters within the data (Li et al. 2022c). The prediction of aerosol pollution episodes and their intensity is made easier by supervised learning techniques like classification algorithms, which learn from labelled data to produce predictions or classify cases into predetermined categories (Gupta et al. 2023). Aerosol pollution and meteorological factors have links that can be revealed using association rule mining tools, which analyse associations or correlations between variables.

3.3 Application of Data Mining in Analysing Aerosol Pollution Data

3.3.1 Clustering Algorithms for Identifying Aerosol Pollution Hotspots

Aerosol concentration patterns provide the basis for several groupings that are formed from the data by clustering techniques like k-means clustering and hierarchical clustering (Zhou et al. 2023). Several studies locate hotspots of increased aerosol pollution levels by clustering spatially resolved aerosol data. This provides important information for focused air quality management activities (Li et al. 2022c). For example, Nakhjiri and Kakroodi (2024) utilize clustering analysis to investigate industrial emissions and air pollution trends in Tehran province, Iran, using Sentinel-5 Precursor data and forecasting models. Similarly, Rendana et al. (2022) classified $PM_{2.5}$ pollution levels in South Sumatra Province using satellite data and hierarchical cluster analysis to support air quality management during forest fire events.

3.3.2 Classification Algorithms for Predicting Aerosol Pollution Events

Decision trees (Tileubai et al. 2023), random forests (Yu et al. 2016), and SVM (Kulkarni et al. 2022) are a few examples of classification algorithms that use labelled data to learn and forecast the likelihood of aerosol pollution events based on emission sources, meteorological variables, and other appropriate information. Certain recent works create predictive models for predicting times of increased aerosol concentrations and determining the possibility of surpassing air quality regulations by training classification algorithms on historical data (Essamlali et al. 2024). Early warning systems and preventative actions to lessen the negative effects of aerosol pollution on the environment and public health are made possible by these models (KELLY et al. 2012).

3.3.3 Association Rule Mining for Identifying Correlations Between Aerosol Pollution and Meteorological Variables

Association rule mining techniques, such as the Apriori algorithm and frequent pattern mining, analyse associations or correlations between aerosol pollution and meteorological variables, revealing underlying relationships and dependencies (Mennis and Liu 2005; Paas et al. 2017; Wu et al. 2024). Historical data on aerosol concentrations and meteorological parameters, several studies uncover patterns such as the influence of wind speed and direction on aerosol dispersion, the impact of temperature inversions on pollutant accumulation, and the relationship between humidity and aerosol hygroscopic growth (Zhou et al. 2015; Chen et al. 2020). These insights inform our understanding of aerosol pollution dynamics and contribute to the development of targeted mitigation strategies.

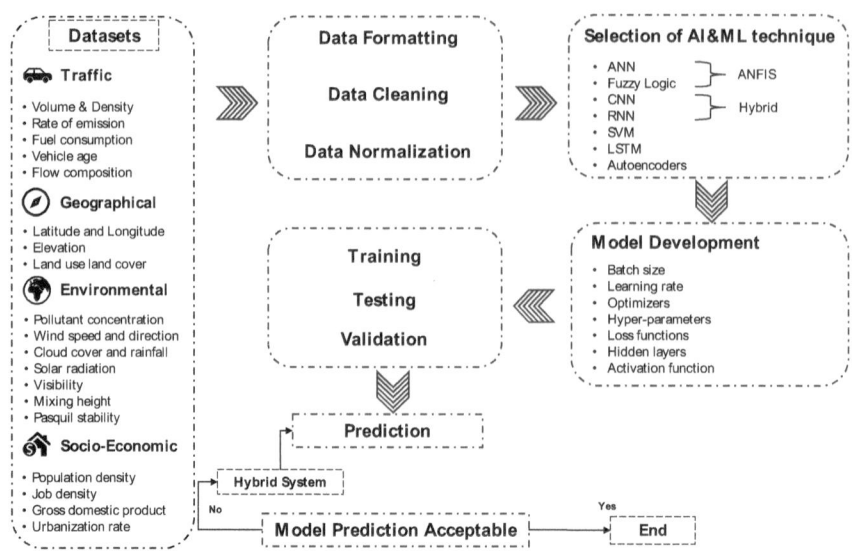

Fig. 3.2 Process of AI&ML-based models for atmospheric environments

3.4 AI&ML Applications in Aerosol Pollution Monitoring and Prediction

3.4.1 Overview of AI&ML Techniques

Aerosol pollution monitoring and prediction have undergone a dramatic change because of the merging of AI&ML techniques (Fig. 3.2) (Tripathi et al. 2024). These innovative methods, complex models that can process large, complex datasets and provide accurate projections have been made easier to create (Rautela et al. 2024a). Aerosol pollution analysis can be accomplished with a variety of capabilities because of the broad array of algorithms and methodologies that make up AI&ML (Gupta et al. 2023). This section examines popular AI&ML methods and explains their uses and functions in aerosol pollution prediction and monitoring.

3.4.2 Application of AI&ML in Aerosol Pollution Monitoring

AI & ML techniques play a crucial role in monitoring aerosol pollution levels, identifying trends, and providing real-time assessments of air quality. By leveraging advanced algorithms and computational methods, several past studies gain significant information about aerosol dynamics and pollution sources by analysing diverse datasets, including ground-based measurements, satellite observations, and atmospheric modelling simulations (Mo et al. 2019; Sun et al. 2019; Javaid et al. 2023).

3.4.3 Time Series Analysis for Predicting Future Aerosol Concentrations

Future aerosol concentration levels are frequently predicted using time series analysis approaches, such as seasonal decomposition and autoregressive integrated moving average (ARIMA) models, based on historical data. Time series analysis involves examining sequences of observations ordered chronologically. It operates on the premise that predicting future values relies on past patterns within the data. Among the plethora of statistical techniques for forecasting variables, stochastic processes stand out for their ability to incorporate uncertainty into predictions. Widely environmental science studies, stochastic models like the Box-Jenkins ARIMA model are instrumental in generating synthetic series mirroring observed data persistence and predicting future trends based on historical data (Wang et al. 2014). Extending beyond hydrology, stochastic processes find utility in climatology domains such as wind speed, precipitation, snow cover, and air and water temperature (Cadenas and Rivera 2010; Kumar et al. 2023). Notably, studies employing ARIMA models have delved into air pollution modelling, forecasting concentrations of pollutants like particulate matter, nitrogen dioxide, sulphur dioxide, carbon monoxide, and ozone (Chattopadhyay and Chattopadhyay 2009). However, in the context of aerosol optical properties, the application of stochastic models remains scarce in India, with existing research focusing mainly on broader regions like the Indo-Gangetic Plain rather than specific locations (Abish and Mohanakumar 2013; Soni et al. 2014).

3.4.4 Neural Networks for Forecasting Aerosol Concentration Levels

Neural networks, inspired by the human brain's neural networks, are powerful computational models capable of learning complex patterns and making predictions based on input data. ANNs emulate the human nervous system, comprising interconnected neurons that collectively address a spectrum of challenges, from function approximation to clustering and optimization (Sofi et al. 2023). The three-stage process involved in ANN modelling, encompassing design, training, and validation, underscores its versatility (Rautela et al. 2022). During the design phase, crucial parameters such as architecture, layers, neurons, and learning algorithms are meticulously chosen (Karagulian et al. 2015). Training involves iterative adjustments of synaptic weights to minimize errors, while validation gauges the network's generalization performance for unknown data. Multilayer Perceptron (MLPs), a prominent type of ANN, have proven effective in predicting atmospheric pollution events. Typically featuring input, hidden, and output layers, MLPs can adapt to complex patterns by incorporating multiple hidden layers. Configuring neurons in the hidden layers is of utmost importance, as an incorrect count can lead to overfitting or under-fitting. Techniques like thumb rule and trial and error, network reduction offer solutions to optimize neuron numbers. The application of ANNs in gaseous pollutants forecasting continued with studies by Slini et al. (2003) and Kandya (2013) both emphasizing the importance of optimizing input parameters for improved accuracy. Comparative assessments with other forecasting techniques consistently positioned ANNs as superior for gaseous pollutants.

3.4.5 Fuzzy Logic (FL) for Air Pollution Prediction

FL, another AI technique, operates on a different paradigm by assigning truth values in a range. Developed from fuzzy set theory, it accommodates linguistic variables, making it adept at handling uncertainty in natural language statements. Fuzzy logic's three main phases—fuzzification, inference, and defuzzification—form a robust modelling system capable of addressing nuanced problems. FL, renowned for its capacity to manage uncertainty, enhanced fault tolerance, and adeptness in handling highly complex nonlinear functions, has garnered extensive adoption in the realm of air pollution prediction. The advantages of FL are exemplified in various studies. For example, Chen et al. (2019) innovatively introduced a novel fuzzy time series model specifically for O_3 prediction, showcasing its superior performance when compared to traditional fuzzy time series models. Jain and Khare (2010) applied a neuro-fuzzy model to predict the concentration of CO in Delhi, achieving accurate estimates at complex urban levels. Carbajal-Hernández et al. (2012) predicts air quality in Mexico City by utilizing the FL model alongside the autoregression model and signal processing. The introduction of a novel algorithm, the "Sigma operator," allowed for precise evaluation of air quality variables, showcasing the effectiveness of fuzzy-based models. Moreover, Al-Shammari (2013), evaluates stochastic and FL-driven models to estimate the daily maximum concentrations of O_3. The findings indicated that the FL-based model exhibited a marginal superiority over the statistical model, particularly in instances of severe pollution events. Innovative approaches like the Fuzzy Inference Ensemble (FIE), as proposed by Bougoudis et al. (2016), demonstrated high accuracy in air pollution forecasting for Athens.

3.4.6 Support Vector Machines (SVM) for Classifying Aerosol Pollution Severity

SVM are supervised learning algorithms that excel in classification tasks, particularly in cases where the data is nonlinearly separable. In aerosol pollution monitoring, SVM models are employed to classify the severity of aerosol pollution events based on input variables such as aerosol concentration levels, meteorological parameters, and emission sources. SVM are popular for supervised learning, excelling in classification, prediction, density estimation, and pattern recognition. SVM seeks an optimal hyperplane to segregate data into predefined classes, with kernel functions playing a pivotal role in introducing nonlinearity. SVM, when combined with other machine learning algorithms, have been helpful in forecasting diverse types of pollutants. Feng et al. (2011) compared SVM with other models for forecasting daily maximum concentrations of O_3 in Beijing, highlighting its stable and accurate performance. Yeganeh et al. (2012) assessed the efficacy of a forecasting model utilizing SVM integrated with Partial Least Squares (PLS) for the prediction of CO concentrations, demonstrating positive outcomes. García Nieto et al. (2013) conducted a comparative analysis of various prediction models for PM_{10} concentrations, determining that the SVM method exhibited superior accuracy and robustness. Luna et al. (2014) utilized Principal PCA in combination with SVM and ANN for the prediction of O_3 levels in Rio de Janeiro. Their study specifically investigated the influence of meteorological parameters on the concentrations of O_3.

3.4.7 Deep Learning Techniques for Image-Based Aerosol Monitoring

Deep Neural Networks (DNNs) represent an advanced version of ANNs, characterized by structural depth and scalability. DNNs, with more than three layers, can automatically extract features from raw inputs, known as feature learning. Notable architectures within DNNs, such as CA, LSTM, CNNs, and RNNs have demonstrated superior performance, especially in air pollution forecasting and pattern identification (Rautela et al. 2024a). The training of DNNs demands significant computational power, leading to advancements in processing capabilities and the development of sophisticated algorithms. Early on, Freeman et al. (2018) employed a combination of Long Short-Term Memory (LSTM) and Recurrent Neural Networks (RNN) to predict ozone concentrations in an urban area. While showing strong predictability in 8-h average ozone concentrations, various model runs revealed overfitting concerns, underscoring the necessity for further refinement. Wang and Song (2018) introduced an ensemble method using a deep LSTM network with fuzzy c-means clustering for air quality forecasting. This ensemble approach outperformed individual models, showcasing its efficacy in both short-term and long-term predictions. Qi et al.

(2019) presented a novel forecasting approach employing a fusion of Graph Convolutional and LSTM (GC-LSTM) neural networks, aiming to investigate spatial interdependence within air quality data. The spatial correlation modelling highlighted the consistency of the GC-LSTM model for short-term forecasting, suggesting potential improvements for long-term predictions with enhanced spatio-temporal considerations. In a novel approach, Li et al. (2015) and Zhang et al. (2016) incorporated large-scale datasets of graphical images for air pollution estimation, utilizing CNN. The models, trained on images capturing various atmospheric conditions, demonstrated improved prediction accuracy, emphasizing the adaptability of deep learning to diverse data types.

3.5 Case Study on Prediction of AARs Using AI&ML

3.5.1 Background

Rapid industrialization, urbanization, and civilization have created severe environmental problems, such as air and soil pollution brought on by industrial emissions and inappropriate waste disposal (Turan et al. 2018). Among these challenges, aerosol pollution has emerged as a critical concern, exacerbated by both natural phenomena like forest fires and human activities such as the combustion of fossil fuels (Ajay et al. 2021). Aerosols, comprising various particles suspended in the atmosphere, play a pivotal role in climate dynamics and environmental health. Aerosol Atmospheric Rivers (AARs), similar to atmospheric rivers (ARs), transport aerosol species across vast distances, significantly impacting air quality and visibility (Chakraborty et al. 2022). Recent studies have highlighted the importance of integrating advanced technologies like satellite remote sensing and ML to monitor and predict aerosol dynamics effectively (Bozdağ et al. 2020; Du et al. 2021; Qiu et al. 2023). ML and DL models, such as CNNs and LSTM, demonstrate promising capabilities in detecting and forecasting aerosol-related variables (Huang and Kuo 2018). However, the black-box nature of some DL models poses challenges in interpreting their predictions, necessitating further exploration of data assimilation techniques to improve model accuracy. Integrating ML and DL models with IAT-based AAR algorithms offers a complementary approach to detecting AARs, enhancing environmental monitoring and forecasting capabilities (Rautela et al. 2024a, b). These models enable accurate predictions in real/near real time, facilitating proactive decision-making to address complex aerosol pollution dynamics and their broader ecological implications.

3.5.2 Methodology

The detected AAR frames of key aerosol species in Chap. 2 are used as an input parameter for the prediction of AARs. The dataset from 2015–2022 on a 6-h time step which comprises 11,684 frames of each aerosol are used as an input dataset to the DL model. The autoencoder network plays a crucial role in predicting the intensity of AARs for various key aerosol species (Rautela et al. 2024a). Comprising three primary components—the encoder, decoder, and bottleneck. The autoencoder architecture undergoes a series of transformations to encode input frames into a latent space representation and decode them back into reconstructed frames. The Encoder utilizes convolutional layers with Rectified Linear Unit (ReLU) activation, batch normalization, and MaxPool2d operations to compress input frames while retaining essential spatial features. Dropout is incorporated for regularization purposes to prevent overfitting. Acting as an information filter, the bottleneck module connects successive input frames by capturing time-related spectral features. On the other hand, the Decoder reconstructs frames by means of "ConvTranspose2d" layers, building successive frames one after the other until the AAR maps for later timestamps are fully rebuilt. Iterative changes for model refinement are made easier by the seamless integration of the reconstruction loss function, or RMSE, with the Adam optimizer.

Batch Normalization (BN) and Activation functions (AF) are essential components of the autoencoder architecture. BN acts as an effective regularization technique, enhancing network stability and efficiency during training by standardizing and normalizing inputs at each layer. It mitigates overfitting and strengthens network robustness by saturating the network with normalization capabilities. Activation functions introduce nonlinearity to capture intricate input–output connections. ReLU and Sigmoid are commonly employed activation functions, with ReLU simplifying computation by transforming negative values to zeros, and Sigmoid being useful for binary classification tasks. The training process uses weight initialization and the Adam optimizer with default settings, giving priority to RMSE reduction for weight updates (Rautela et al. 2024b).

The model's hyperparameters consist of L_2 loss for frame reconstruction, Adam optimization in gradient descent, and a batch size of 1 for every 10 frames in each iteration (Singh and Goyal 2023a). The input data consists of frames showing the intensity of identified AARs for five distinct species, and the training loss function is RMSE. The model's learning progress is demonstrated by the training and testing loss changes throughout epochs (Fig. 3.3). The autoencoder network, when combined with BN and AF components, creates a strong framework for estimating AAR intensity, enabling improved comprehension and control of the dynamics of aerosol pollution.

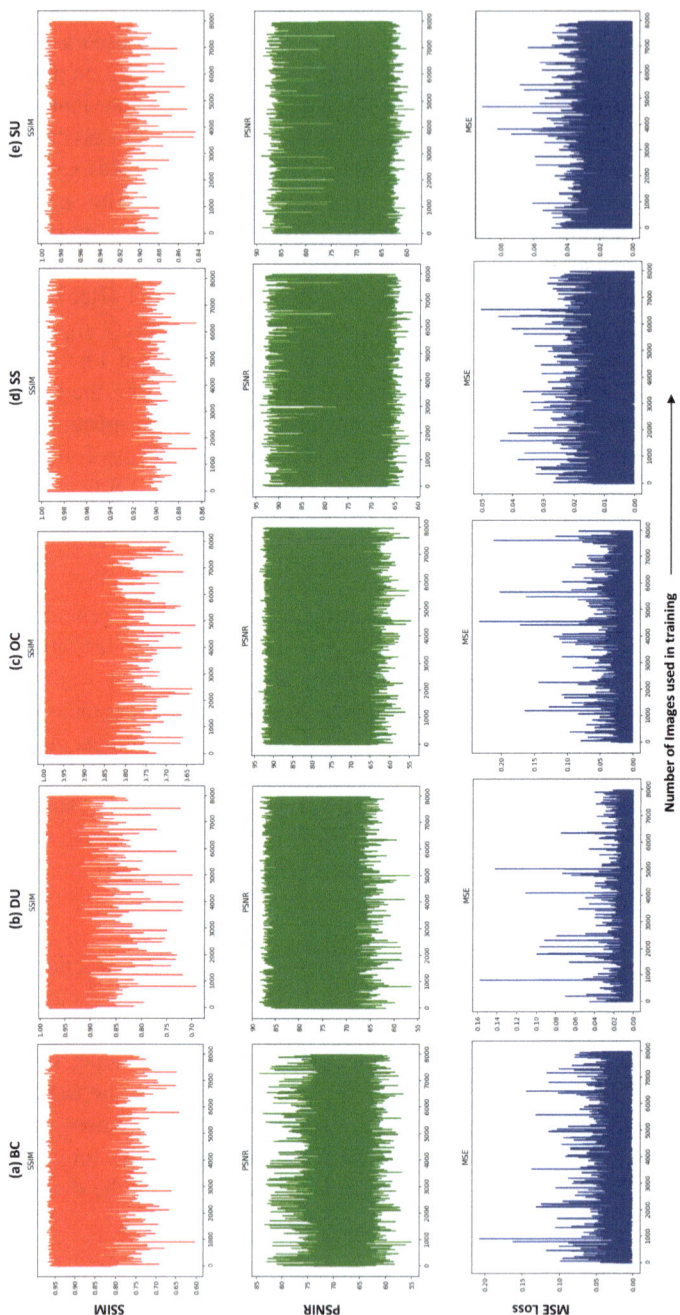

Fig. 3.3 Model evaluation parameters for key aerosol species in the training phase

3.5.3 Results

The development and evaluation of five autoencoder models for predicting extreme aerosol events were conducted, with datasets divided in the ratio of 70%, 20%, and 10% to training, testing, and validation sets, respectively. Each model was trained up to 30 epochs for each aerosol species. Performance evaluation was carried out using established metrics such as structural similarity index (SSIM) and peak signal-to-noise ratio (PSNR). These metrics provided insights into the predictive capabilities of the models, with SSIM assessing image similarity and PSNR indicating the quality of image reconstruction (Meghani et al. 2023). Additionally, machine learning metrics, including root mean square error (RMSE) and mean square error (MSE), were employed to quantify dissimilarity between predicted and actual values.

The outcomes of the model evaluation showed how well the autoencoder models captured complex data relationships and made accurate predictions. During the training phase, the SSIM values ranged from 0.85 to 0.96, indicating a high degree of similarity between the predicted and actual images. Similarly, PSNR values varied from 1.26 to 41.26 dB, where higher values indicate better quality image reconstruction. Furthermore, the models were able to forecast aerosol intensity with good accuracy; RMSE values ranged from 1.59 to 17.02 mg/(m-sec), while MSE values were within the same range, 2.52 to 289.60 mg/(m-sec).

Comparing metrics scores between training, testing, and validation sets revealed consistent patterns across all phases, indicating the robustness of the models' performance. Together with reduced RMSE and MSE values, the greater SSIM and PSNR values highlighted the accuracy and calibre of the models' predictions. Notably, the achieved metrics scores surpassed benchmarks set by previous studies, attributed to the autoencoder models' effectiveness in capturing complex spatio-temporal features within the aerosol data.

Several factors contributed to the superior performance of the autoencoder models. The models' ability to identify important patterns in the data was improved by the thoughtful application of several layers and methods, such as Batch Normalization, Conv2d, and Up sampling. Additionally, the models' ability to effectively capture complex relationships inherent in the aerosol data played a pivotal role in improving prediction accuracy. Moreover, the models' efficacy in reconstructing aerosol intensity maps was demonstrated visually through Fig. 3.4, showcasing its prediction and training capacities. Sequences of 10 input AAR frame of BC AARs and predicts the 11th AAR frame and further compare with the actual one, and the model's predicted output for the same time step were shown in these samples (Fig. 3.4a). Also, the predicted and observed AAR frames for DU, OC, SS, and SU AARs are shown in the Fig. 3.4b, where the models showed outstanding predicted accuracy, even though some plots in the sample sets were blank because associated aerosol atmospheric rivers were not available at time steps.

(a)

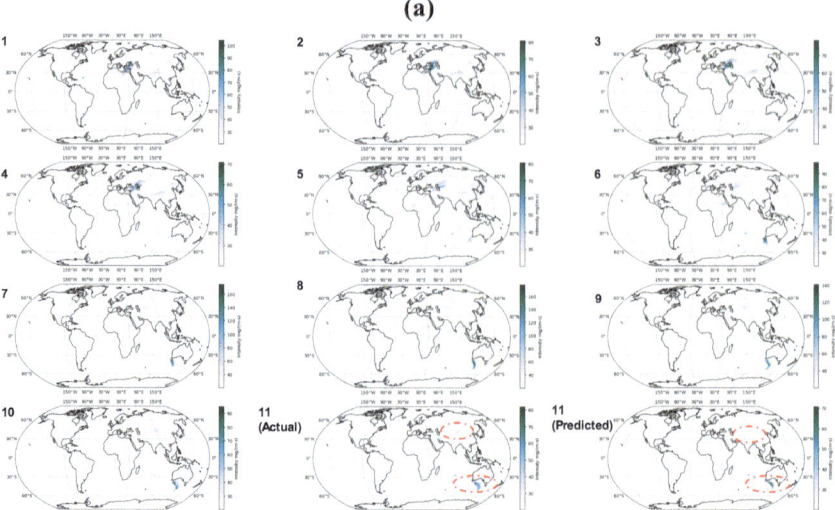

Fig. 3.4 **a** Patterns of 10 sequential BC AARs frames used as an input to predict the 11th frame and compare with the actual frame **b** other predicted key aerosol species frames with comparison with the actual ones

3.6 Challenges and Future Directions

Addressing data quality and availability challenges remains a critical obstacle in aerosol pollution analysis using data mining and AI&ML techniques. Predictive model development and validation are hampered by the lack of access to complete and high-quality datasets, such as satellite observations, ground-based measurements, and atmospheric modelling outputs. In addition, there are considerable logistical and technical difficulties in combining various data sources, necessitating the use of reliable data integration frameworks and compatible data formats. There are additional difficulties in interpreting and understanding AI&ML models, especially in complex and nonlinear systems like aerosol pollution, where model interpretability and explainability are essential for making well-informed decisions (Fan et al. 2017; Qiu et al. 2023; Singh and Goyal 2023b). Developing advanced data integration techniques to seamlessly combine diverse datasets, enhancing model interpretability and transparency through novel visualization and explainability techniques, and improving data quality and accessibility through improved data collection methods and data sharing initiatives are some of the future research directions in aerosol pollution analysis using data mining and AI&ML techniques. For more precise and reliable forecasts of aerosol pollution dynamics and their effects on the environment and public health, upcoming studies should also concentrate on investigating cutting-edge AI&ML approaches, such as deep learning and reinforcement learning.

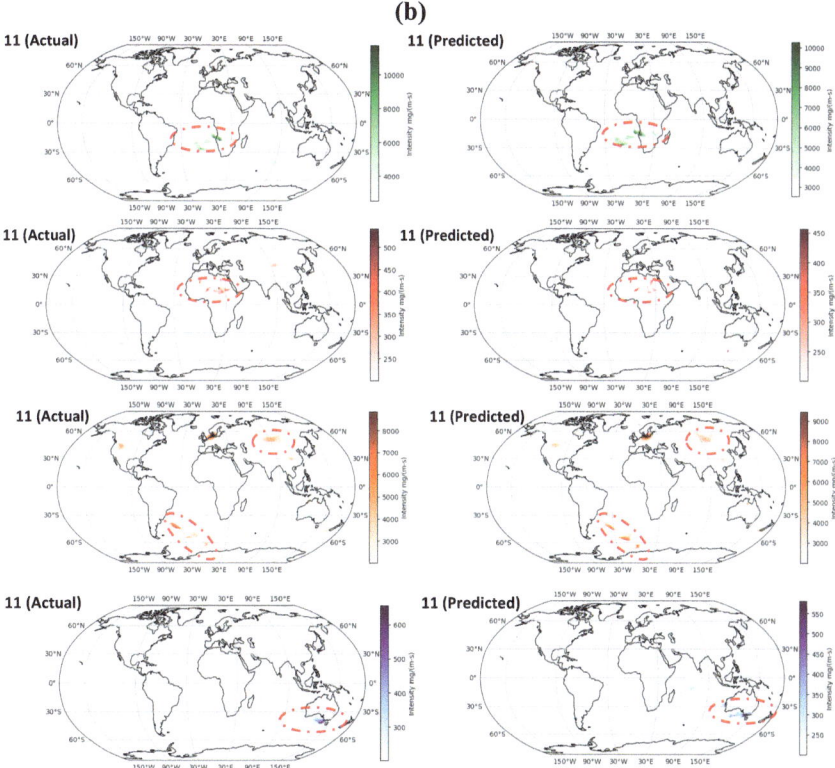

Fig. 3.4 (continued)

Our understanding of aerosol pollution may be improved, and evidence-based policies for air quality management and public health protection can be informed by addressing these issues and promoting research in these areas.

3.7 Conclusion

The combination of data mining and AI&ML approaches is a major step forward in tackling the complex relationship related to aerosol pollution and atmospheric dynamics. These sophisticated analytical techniques have a wide range of applications in aerosol science that present unprecedented opportunities to enhance our understanding of aerosol dynamics, sources of pollution, and their effects on human and environmental health. Aerosol concentration patterns may be found in space and time using data mining techniques including clustering, classification, and association rule mining. This helps identify underlying trends and causes that contribute to aerosol pollution. Deep learning models, neural networks, fuzzy logic, SVM, and other AI&ML approaches make it easier to create predictive models for atmospheric interactions, aerosol concentrations, and dispersion patterns. These models provide precise estimates and evaluations of aerosol pollution levels by utilizing satellite imagery, atmospheric simulations, and observational data.

Applications and case studies demonstrate the usefulness of data mining and AI&ML methods in solving real-world aerosol pollution problems. The case studies show how sophisticated analytical techniques can improve environmental monitoring and forecasting, from locating aerosol pollution hotspots over the globe to forecasting aerosol atmospheric rivers. For extreme aerosol episodes, the autoencoder models created in this chapter demonstrated strong prediction accuracy. Their effectiveness in collecting complicated spatio-temporal patterns in aerosol data was demonstrated by their higher performance metrics scores and ability to predict aerosol intensity properly. The results have important ramifications for companies depending on precise predictions of aerosol pollution events and advance aerosol science's predictive modelling. There are still issues with data availability, quality, and model interpretability even with the tremendous advancements made. To support better-informed decision-making in aerosol pollution control, future research should concentrate on developing model interpretability approaches, strengthening data integration frameworks, and improving data collection methodologies. The integration of data mining and AI&ML techniques offers a paradigm shift in how we approach the study of aerosol pollution and atmospheric dynamics. By harnessing the power of advanced analytical tools, researchers can gain deeper insights into aerosol dynamics, improve air quality management strategies, and ultimately safeguard human health and the environment from the adverse impacts of aerosol pollution.

References

Abish B, Mohanakumar K (2013) A stochastic model for predicting aerosol optical depth over the north Indian region. Int J Remote Sens 34:1449–1458. https://doi.org/10.1080/01431161.2012.723149

Ajay A, Krishna Moorthy K, Satheesh SK, Ilavazhagan G (2021) Impact assessment of change in anthropogenic emissions due to lockdown on aerosol characteristics in a rural location. Curr Sci 120:332–340. https://doi.org/10.18520/cs/v120/i2/332-340

Al-Shammari ET (2013) Public warning systems for forecasting ambient ozone pollution in Kuwait. Environ Syst Res 2:2. https://doi.org/10.1186/2193-2697-2-2

Alzubaidi L, Zhang J, Humaidi AJ et al (2021) Review of deep learning: concepts, CNN architectures, challenges, applications, future directions. J Big Data 8:53. https://doi.org/10.1186/s40537-021-00444-8

Bellinger C, Mohomed Jabbar MS, Zaïane O, Osornio-Vargas A (2017) A systematic review of data mining and machine learning for air pollution epidemiology. BMC Public Health 17:907. https://doi.org/10.1186/s12889-017-4914-3

Bin Tarek MF, Asaduzzaman M, Patwary M (2018) Spatio-temporal analysis of large air pollution data. In: 2018 10th international conference on electrical and computer engineering (ICECE). IEEE, pp 221–224

Bougoudis I, Demertzis K, Iliadis L (2016) HISYCOL a hybrid computational intelligence system for combined machine learning: the case of air pollution modeling in Athens. Neural Comput Appl 27:1191–1206. https://doi.org/10.1007/s00521-015-1927-7

Bozdağ A, Dokuz Y, Gökçek ÖB (2020) Spatial prediction of PM10 concentration using machine learning algorithms in Ankara. Turkey. Environ Pollut 263:114635. https://doi.org/10.1016/j.envpol.2020.114635

Brun J, Shrestha P, Barros AP (2011) Mapping aerosol intrusion in Himalayan valleys using the moderate resolution imaging Spectroradiometer (MODIS) and cloud-aerosol lidar and infrared pathfinder satellite observation (CALIPSO). Atmos Environ 45:6382–6392. https://doi.org/10.1016/j.atmosenv.2011.08.026

Cadenas E, Rivera W (2010) Wind speed forecasting in three different regions of Mexico, using a hybrid ARIMA–ANN model. Renew Energy 35:2732–2738. https://doi.org/10.1016/j.renene.2010.04.022

Carbajal-Hernández JJ, Sánchez-Fernández LP, Carrasco-Ochoa JA, Martínez-Trinidad JF (2012) Assessment and prediction of air quality using fuzzy logic and autoregressive models. Atmos Environ 60:37–50. https://doi.org/10.1016/j.atmosenv.2012.06.004

Chakraborty S, Guan B, Waliser DE, da Silva AM (2022) Aerosol atmospheric rivers: climatology, event characteristics, and detection algorithm sensitivities. Atmos Chem Phys 22:8175–8195. https://doi.org/10.5194/acp-22-8175-2022

Chattopadhyay G, Chattopadhyay S (2009) Autoregressive forecast of monthly total ozone concentration: a neurocomputing approach. Comput Geosci 35:1925–1932. https://doi.org/10.1016/j.cageo.2008.11.007

Chen J, de Hoogh K, Gulliver J et al (2019) A comparison of linear regression, regularization, and machine learning algorithms to develop Europe-wide spatial models of fine particles and nitrogen dioxide. Environ Int 130:104934. https://doi.org/10.1016/j.envint.2019.104934

Chen Z, Chen D, Zhao C, et al (2020) Influence of meteorological conditions on PM2.5 concentrations across China: A review of methodology and mechanism. Environ Int 139:105558. https://doi.org/10.1016/j.envint.2020.105558

Chowdhury A-S, Uddin MS, Tanjim MR, et al (2020) Application of data mining techniques on air pollution of Dhaka City. In: 2020 IEEE 10th international conference on intelligent systems (IS). IEEE, pp 562–567

Christensen MW, Jones WK, Stier P (2020) Aerosols enhance cloud lifetime and brightness along the stratus-to-cumulus transition. Proc Natl Acad Sci 117:17591–17598. https://doi.org/10.1073/pnas.1921231117

Du S, Li T, Yang Y, Horng S-J (2021) Deep air quality forecasting using hybrid deep learning framework. IEEE Trans Knowl Data Eng 33:2412–2424. https://doi.org/10.1109/TKDE.2019.2954510

Elahi M, Afolaranmi SO, Martinez Lastra JL, Perez Garcia JA (2023) A comprehensive literature review of the applications of AI techniques through the lifecycle of industrial equipment. Discov Artif Intell 3:43. https://doi.org/10.1007/s44163-023-00089-x

Essamlali I, Nhaila H, El Khaili M (2024) Supervised machine learning approaches for predicting key pollutants and for the sustainable enhancement of urban air quality: a systematic review. Sustainability 16:976. https://doi.org/10.3390/su16030976

Fan J, Li Q, Hou J, et al (2017) A spatiotemporal prediction framework for air pollution based on deep RNN. ISPRS Ann Photogramm Remote Sens Spat Inf Sci IV-4/W2:15–22. https://doi.org/10.5194/isprs-annals-IV-4-W2-15-2017

Feng Y, Zhang W, Sun D, Zhang L (2011) Ozone concentration forecast method based on genetic algorithm optimized back propagation neural networks and support vector machine data classification. Atmos Environ 45:1979–1985. https://doi.org/10.1016/j.atmosenv.2011.01.022

Filonchyk M, Yan H, Zhang Z et al (2019) Combined use of satellite and surface observations to study aerosol optical depth in different regions of China. Sci Rep 9:6174. https://doi.org/10.1038/s41598-019-42466-6

García Nieto PJ, Combarro EF, del Coz Díaz JJ, Montañés E (2013) A SVM-based regression model to study the air quality at local scale in Oviedo urban area (Northern Spain): a case study. Appl Math Comput 219:8923–8937. https://doi.org/10.1016/j.amc.2013.03.018

Gibert K, Izquierdo J, Sànchez-Marrè M et al (2018) Which method to use? An assessment of data mining methods in environmental data science. Environ Model Softw 110:3–27. https://doi.org/10.1016/j.envsoft.2018.09.021

Goyal MK (2014) Modeling of sediment yield prediction using M5 model tree algorithm and wavelet regression. Water Resour Manag 28:1991–2003. https://doi.org/10.1007/s11269-014-0590-6

Goyal MK, Burn DH, Ojha CSP (2012) Evaluation of machine learning tools as a statistical downscaling tool: temperatures projections for multi-stations for Thames River Basin, Canada. Theor Appl Climatol 108:519–534. https://doi.org/10.1007/s00704-011-0546-1

Goyal MK, Ojha CSP (2011) Evaluation of linear regression methods as downscaling tools in temperature projections over the Pichola Lake Basin in India. Hydrol Process 25:1453–1465. https://doi.org/10.1002/hyp.7911

Goyal MK, Sharma A (2016) A fuzzy c-means approach regionalization for analysis of meteorological drought homogeneous regions in western India. Nat Hazards 84:1831–1847. https://doi.org/10.1007/s11069-016-2520-9

Gupta NS, Mohta Y, Heda K et al (2023) Prediction of air quality index using machine learning techniques: a comparative analysis. J Environ Public Health 2023:1–26. https://doi.org/10.1155/2023/4916267

Haque M, Singh R (2017) Air pollution and human health in Kolkata, India: a case study. Climate 5:77. https://doi.org/10.3390/cli5040077

Hu M, Wang Y, Wang S et al (2021) Spatial-temporal heterogeneity of air pollution and its relationship with meteorological factors in the Pearl River Delta. China Atmos Environ 254:118415. https://doi.org/10.1016/j.atmosenv.2021.118415

Huang C-J, Kuo P-H (2018) A Deep CNN-LSTM model for particulate matter (PM2.5) forecasting in smart cities. Sensors 18:2220. https://doi.org/10.3390/s18072220

Jain S, Khare M (2010) Adaptive neuro-fuzzy modeling for prediction of ambient CO concentration at urban intersections and roadways. Air Qual Atmos Heal 3:203–212. https://doi.org/10.1007/s11869-010-0073-8

Javaid M, Haleem A, Khan IH, Suman R (2023) Understanding the potential applications of artificial intelligence in agriculture sector. Adv Agrochem 2:15–30. https://doi.org/10.1016/j.aac.2022.10.001

Kandya A (2013) Forecasting the tropospheric ozone using artificial neural network modelling approach: a case study of megacity Madras, India. J Civ Environ Eng 1.https://doi.org/10.4172/2165-784x.s1-006

Karagulian F, Belis CA, Dora CFC et al (2015) Contributions to cities' ambient particulate matter (PM): a systematic review of local source contributions at global level. Atmos Environ 120:475–483. https://doi.org/10.1016/j.atmosenv.2015.08.087

Kelly FJ, Fuller GW, Walton HA, Fussell JC (2012) Monitoring air pollution: use of early warning systems for public health. Respirology 17:7–19. https://doi.org/10.1111/j.1440-1843.2011.02065.x

Kotu V, Deshpande B (2015) Data mining process. In: Predictive analytics and data mining. Elsevier, pp 17–36

Krishan M, Jha S, Das J et al (2019) Air quality modelling using long short-term memory (LSTM) over NCT-Delhi, India. Air Qual Atmos Heal 12:899–908. https://doi.org/10.1007/s11869-019-00696-7

Kulkarni M, Raut A, Chavan S, et al (2022) Air quality monitoring and prediction using SVM. In: 2022 6th international conference on computing, communication, control and automation (ICCUBEA. IEEE, pp 1–4

Kulmala M, Kokkonen T, Ezhova E et al (2023) Aerosols, clusters, greenhouse gases, trace gases and boundary-layer dynamics: on feedbacks and interactions. Boundary-Layer Meteorol 186:475–503. https://doi.org/10.1007/s10546-022-00769-8

Kumar M, Tiwari RK, Kumar K, Rautela KS (2023) Statistical evaluation of snow accumulation and depletion from remotely sensed MODIS snow time series data using the SARIMA model. J Water Supply Res Technol. https://doi.org/10.2166/aqua.2023.231

Kuniyal JC, Guleria RP (2019) The current state of aerosol-radiation interactions: a mini review. J Aerosol Sci 130:45–54. https://doi.org/10.1016/j.jaerosci.2018.12.010

Li H, Yang Y, Wang H et al (2022a) Projected aerosol changes driven by emissions and climate change using a machine learning method. Environ Sci Technol 56:3884–3893. https://doi.org/10.1021/acs.est.1c04380

Li J, Carlson BE, Yung YL et al (2022b) Scattering and absorbing aerosols in the climate system. Nat Rev Earth Environ 3:363–379. https://doi.org/10.1038/s43017-022-00296-7

Li J, Hendricks J, Righi M, Beer CG (2022c) An aerosol classification scheme for global simulations using the K-means machine learning method. Geosci Model Dev 15:509–533. https://doi.org/10.5194/gmd-15-509-2022

Liu D, He C, Schwarz JP, Wang X (2020) Lifecycle of light-absorbing carbonaceous aerosols in the atmosphere. npj Clim Atmos Sci 3:40. https://doi.org/10.1038/s41612-020-00145-8

Luna AS, Paredes MLL, de Oliveira GCG, Corrêa SM (2014) Prediction of ozone concentration in tropospheric levels using artificial neural networks and support vector machine at Rio de Janeiro, Brazil. Atmos Environ 98:98–104. https://doi.org/10.1016/j.atmosenv.2014.08.060

Ma D, Gao J, Zhang Z, Zhao H (2021) Identifying atmospheric pollutant sources using a machine learning dispersion model and Markov chain Monte Carlo methods. Stoch Environ Res Risk Assess 35:271–286. https://doi.org/10.1007/s00477-021-01973-7

Masoud AA (2023) Spatio-temporal patterns and trends of the air pollution integrating MERRA-2 and in situ air quality data over Egypt (2013–2021). Air Qual Atmos Heal 16:1543–1570. https://doi.org/10.1007/s11869-023-01357-6

Meghani S, Singh S, Kumar N, Goyal MK (2023) Predicting the spatiotemporal characteristics of atmospheric rivers: a novel data-driven approach. Glob Planet Change 231:104295. https://doi.org/10.1016/j.gloplacha.2023.104295

Mennis J, Liu JW (2005) Mining association rules in spatio-temporal data: an analysis of urban socioeconomic and land cover change. Trans GIS 9:5–17. https://doi.org/10.1111/j.1467-9671.2005.00202.x

Mo, Zhang, Li, Qu (2019) A Novel air quality early-warning system based on artificial intelligence. Int J Environ Res Public Health 16:3505.https://doi.org/10.3390/ijerph16193505

Nakhjiri A, Kakroodi AA (2024) Air pollution in industrial clusters: a comprehensive analysis and prediction using multi-source data. Ecol Inform 80:102504. https://doi.org/10.1016/j.ecoinf.2024.102504

Nandakumar A (2022) Classification of air pollution levels using supervised machine learning algorithm 10:523–529

Oh H-J, Ma Y, Kim J (2020) Human inhalation exposure to aerosol and health effect: aerosol monitoring and modelling regional deposited doses. Int J Environ Res Public Health 17:1923. https://doi.org/10.3390/ijerph17061923

Ojha CSP, Goyal MK, Kumar S (2007) Applying fuzzy logic and the point count system to select landfill sites. Environ Monit Assess 135:99–106. https://doi.org/10.1007/s10661-007-9713-3

Paas B, Stienen J, Vorländer M, Schneider C (2017) Modelling of urban near-road atmospheric pm concentrations using an artificial neural network approach with acoustic data input. Environments 4:26. https://doi.org/10.3390/environments4020026

Peng Z, Zhang B, Wang D et al (2024) Application of machine learning in atmospheric pollution research: a state-of-art review. Sci Total Environ 910:168588. https://doi.org/10.1016/j.scitotenv.2023.168588

Prank M, Sofiev M, Tsyro S et al (2016) Evaluation of the performance of four chemical transport models in predicting the aerosol chemical composition in Europe in 2005. Atmos Chem Phys 16:6041–6070. https://doi.org/10.5194/acp-16-6041-2016

Qiu M, Zigler C, Selin NE (2022) Statistical and machine learning methods for evaluating trends in air quality under changing meteorological conditions. Atmos Chem Phys 22:10551–10566. https://doi.org/10.5194/acp-22-10551-2022

Qiu Y, Feng J, Zhang Z, et al (2023) Regional aerosol forecasts based on deep learning and numerical weather prediction. npj Clim Atmos Sci 6:71. https://doi.org/10.1038/s41612-023-00397-0

Rautela KS, Kumar D, Gandhi BGR, et al (2022) Application of ANNs for the modeling of streamflow, sediment transport, and erosion rate of a high-altitude river system in Western Himalaya, Uttarakhand. RBRH 27.https://doi.org/10.1590/2318-0331.272220220045

Rautela KS, Singh S, Goyal MK (2024a) Characterizing the spatio-temporal distribution, detection, and prediction of aerosol atmospheric rivers on a global scale. J Environ Manage 351:119675. https://doi.org/10.1016/j.jenvman.2023.119675

Rautela KS, Singh S, Goyal MK (2024b) Resilience to air pollution: A novel approach for detecting and predicting aerosol atmospheric rivers within earth system boundaries earth systems and environment. https://doi.org/10.1007/s41748-024-00421-0

Rendana M, Razi Idris WM, Abdul Rahim S (2022) Clustering analysis of PM2.5 concentrations in the South Sumatra Province, Indonesia, using the Merra-2 satellite application and hierarchical cluster method. AIMS Environ Sci 9:754–770. https://doi.org/10.3934/environsci.2022043

Rojas M, Quintero F, Young N (2020) Analysis of stage-discharge relationship stability based on historical ratings. Hydrology 7:31. https://doi.org/10.3390/hydrology7020031

Sahak N, Asmat A, Hazali NA et al (2019) Multiangle Imaging Spectroradiometer (MISR) and moderate resolution imaging spectrometer (MODIS) aerosol optical depth (AOD) spatial variations in Peninsular Malaysia. IOP Conf Ser Earth Environ Sci 373:012010. https://doi.org/10.1088/1755-1315/373/1/012010

Sekar C, Gurjar BR, Ojha CSP, Goyal MK (2016) Potential assessment of neural network and decision tree algorithms for forecasting ambient PM2.5 and CO concentrations: case study. J Hazardous, Toxic, Radioact Waste 20:. https://doi.org/10.1061/(ASCE)HZ.2153-5515.0000276

Senthil Kumar AR, Goyal MK, Ojha CSP et al (2013) Application of artificial neural network, fuzzy logic and decision tree algorithms for modelling of streamflow at Kasol in India. Water Sci Technol 68:2521–2526.https://doi.org/10.2166/wst.2013.491

Shakya D, Deshpande V, Goyal MK, Agarwal M (2023) PM2.5 air pollution prediction through deep learning using meteorological, vehicular, and emission data: a case study of New Delhi, India. J Clean Prod 427:139278. https://doi.org/10.1016/j.jclepro.2023.139278

Singh S, Goyal MK (2023a) An innovative approach to predict atmospheric rivers: exploring convolutional autoencoder. Atmos Res 289:106754. https://doi.org/10.1016/j.atmosres.2023.106754

Singh S, Goyal MK (2023b) Enhancing climate resilience in businesses: the role of artificial intelligence. J Clean Prod 418:138228. https://doi.org/10.1016/j.jclepro.2023.138228

Slini T, Karatzas K, Moussiopoulos N (2003) Correlation of air pollution and meteorological data using neural networks. Int J Environ Pollut 20:218. https://doi.org/10.1504/IJEP.2003.004279

Sofi MS, Rautela KS, Muslim M et al (2023) Modeling the hydrological response of a snow-fed river in the Kashmir Himalayas through SWAT and artificial neural network. Int J Environ Sci Technol. https://doi.org/10.1007/s13762-023-05170-7

Soni K, Kapoor S, Parmar KS, Kaskaoutis DG (2014) Statistical analysis of aerosols over the Gangetic-Himalayan region using ARIMA model based on long-term MODIS observations. Atmos Res 149:174–192. https://doi.org/10.1016/j.atmosres.2014.05.025

Sun J, Di L, Sun Z et al (2019) County-level soybean yield prediction using deep CNN-LSTM model. Sensors 19:4363. https://doi.org/10.3390/s19204363

Tileubai A, Tsend J, Oyunbileg BE et al (2023) Study of decision tree algorithms: effects of air pollution on under five mortality in Ulaanbaatar. BMJ Heal Care Informatics 30:1–5. https://doi.org/10.1136/bmjhci-2022-100678

Tripathi AK, Aruna M, Parida S et al (2024) Integrated smart dust monitoring and prediction system for surface mine sites using IoT and machine learning techniques. Sci Rep 14:7587. https://doi.org/10.1038/s41598-024-58021-x

Turan V, Khan SA, Mahmood-ur-Rahman, et al (2018) Promoting the productivity and quality of brinjal aligned with heavy metals immobilization in a wastewater irrigated heavy metal polluted soil with biochar and chitosan. Ecotoxicol Environ Saf 161:409–419.https://doi.org/10.1016/j.ecoenv.2018.05.082

Wang HR, Wang C, Lin X, Kang J (2014) An improved ARIMA model for precipitation simulations. Nonlinear Process Geophys 21:1159–1168. https://doi.org/10.5194/npg-21-1159-2014

Wu X, Wen Q, Zhu J (2024) Association rule mining with a special rule coding and dynamic genetic algorithm for air quality impact factors in Beijing. China. Plos One 19:e0299865. https://doi.org/10.1371/journal.pone.0299865

Yeganeh B, Motlagh MSP, Rashidi Y, Kamalan H (2012) Prediction of CO concentrations based on a hybrid partial least square and support vector machine model. Atmos Environ 55:357–365. https://doi.org/10.1016/j.atmosenv.2012.02.092

Yu R, Yang Y, Yang L, et al (2016) RAQ-A random forest approach for predicting air quality in urban sensing systems. Sensors (Basel) 16.https://doi.org/10.3390/s16010086

Yu Y, Si X, Hu C, Zhang J (2019) A review of recurrent neural networks: LSTM cells and network architectures. Neural Comput 31:1235–1270. https://doi.org/10.1162/neco_a_01199

Zhang L, Yang G, Li X (2020) Mining sequential patterns of PM2.5 pollution between 338 cities in China. J Environ Manage 262:110341. https://doi.org/10.1016/j.jenvman.2020.110341

Zhou W, Tie X, Zhou G, Liang P (2015) Possible effects of climate change of wind on aerosol variation during winter in Shanghai, China. Particuology 20:80–88. https://doi.org/10.1016/j.partic.2014.08.008

Zhou Y, Wu T, Zhou Y et al (2023) Can global warming bring more dust? Clim Dyn 61:2693–2715. https://doi.org/10.1007/s00382-023-06706-w

Chapter 4
Aerosol Atmospheric Rivers: Impacts on Particulate Matter Concentrations and Risk Assessment

Abstract Air pollution, particularly from particulate matter (PM), has become a critical global issue, impacting human health and the environment. Aerosol Optical Depth (AOD) serves as a proxy for particulate matter (PM) concentrations in the atmosphere, with higher AOD values generally indicating higher PM levels. AOD is estimated using satellite remote sensing techniques that measure the attenuation of solar radiation caused by aerosols in the atmosphere. Satellite-based data for AOD estimation is necessary due to its ability to provide global coverage and frequent observations, which are essential for monitoring air quality on a large scale. However, empirical methods are also useful to compute $PM_{2.5}$ data from satellite images. This chapter explores the possibility of empirical models to compute the $PM_{2.5}$ concentration and correlate it with the observed datasets as a case study. There is a very strong correlation ($r = 0.75$) between the $PM_{2.5 \text{ satellite data}}$ and $PM_{2.5 \text{ observed data}}$. Furthermore, the presence of Aerosol Atmospheric Rivers (AARs) significantly affects the $PM_{2.5}$ concentrations at that geographic location. However, the risk analysis shows cities like New Delhi, Lahore, Dhaka, and Dammam are at extreme risk of air pollution followed by Ghaziabad, Chongqing, Kolkata, Mumbai, and East London at high high-risk. These findings underscore the urgent need for targeted interventions to mitigate the adverse impacts of air pollution, especially in densely populated and industrialized regions.

Keywords Air pollution · Aerosol Atmospheric Rivers (AARs) · Particulate matter (PM) · Reanalysis datasets · Satellite datasets

4.1 Introduction

Over recent decades, the escalating issue of air pollution has garnered significant global attention, driven by numerous acute pollution episodes observed in various regions worldwide (Shakya et al. 2023). Countries such as China, India, and numerous other developing and low-income nations have been threatened by severe air quality challenges, resulting from rapid industrialization and urban expansion

(Anwar et al. 2021). Specifically, metropolitan hubs have grappled with alarming air pollution levels, particularly exacerbated during the winter months (Krishan et al. 2019). Additionally, several state capitals across India remain highly susceptible to air pollution crises. The Great London Smog of 1952 and the extreme air pollution incidents in Donora, USA in the late 1940s were significant events that contributed to the statistical evidence of the negative health impacts of excessive air pollution (Jacobs et al. 2018). Balakrishnan et al. (2019) highlighted air pollution-related causes were responsible for 1.24 million fatalities in India in 2017 alone.

Particulate matter (PM) exerts detrimental effects not only on living organisms, but also on various climatic parameters, such as temperature, cloud formation, and fog density. The World Health Organization (WHO) reports approximately 4.2 million people die from ambient air pollution-related causes annually (WHO 2021). Additionally, an estimated 7 million deaths are attributed specifically to PM air pollution. This results in an economic loss of $2.9 trillion, which accounts for about 3.3% of the world's GDP (WHO 2021). The presence of fine particles, specifically $PM_{2.5}$ and PM_{10}, in the atmosphere significantly impacts human health, with the ability to infiltrate the lungs and contribute to respiratory illness, cardiovascular disease, and cerebrovascular disease (Fuller et al. 2022). According to Li et al. (2023), there has been a notable escalation in global population-weighted (PW) $PM_{2.5}$ exposure, rising from 28.3 $\mu g/m^3$ in 1998 to a peak of 38.9 $\mu g/m^3$ in 2011. Subsequently, there has been a gradual decline to 34.7 $\mu g/m^3$ by 2019. This surge in $PM_{2.5}$ concentrations has corresponded with an uptick in extreme pollution occurrences, thereby heightening associated health hazards.

Furthermore, as discussed in the previous chapters, aerosol concentrations serve as a crucial indicator for estimating level PM concentrations (Chakraborty et al. 2022; Rautela et al. 2024a). Therefore in the lower tropospheric region, there is an urgent need for continuous and real/near real-time monitoring of aerosols, spanning from large-scale to small-scale observations, to inform various planning and implementation efforts (Guleria and Chand 2020; Chand et al. 2021). Numerous ground-based projects and programmes dedicated to monitoring Aerosol Optical Depth (AOD) are operational worldwide. Examples include AERONET (AErosol RObotic NETwork), Multi-Filter Rotating Shadow band Radiometer (MFRSR), Meteorological Administration Aerosol Remote Sensing Network (CASRNET), Maritime Aerosol Network (MAN), China, etc. (Ranjan et al. 2021). These initiatives provide near real-time AOD data from various network stations globally. However, the geographical coverage of ground monitoring stations remains limited in many countries, underscoring the importance of satellite-based monitoring systems for obtaining spatial data on a broader scale (Anggraini et al. 2024).

Satellite remote sensing has emerged as a powerful tool for monitoring ambient air quality, offering advantages such as high spatio-temporal resolution (Rautela et al. 2024b). The limitations of ground-based air quality monitoring systems, particularly in relation to PM and AOD, may be addressed by satellite remote sensing. For example, the PM concentration can be found using MERRA-2 satellite reanalysis datasets (Randles et al. 2017). Satellite remote sensing-based air quality monitoring holds great potential, but it is still in the research stage with issues to be resolved in establishing its accuracy, consistency, and dependability (Xian et al. 2013). Consequently, the efforts are focused on developing more accurate and reliable methods for air quality monitoring, with particular emphasis on estimating AOD and PM concentrations.

The presence of Aerosol Atmospheric Rivers (AARs) significantly contributes to the escalation of particulate matter (PM) concentrations across regions, consequently amplifying the air quality index of affected areas. This increase in PM levels poses harmful impacts on both human health and the environment, thereby exerting adverse effects on the economy as well. AARs play a crucial role in facilitating the transportation of $PM_{2.5}$ from one geographical region to another. For example, notable instances such as the Godzilla dust storm in 2020 transported dust aerosols from the Sahara Desert to the Caribbean islands and the USA, leading to a substantial deterioration in air quality within the affected regions. Similarly, the Indian dust storm in 2018 resulted in a significant increase in PM concentrations across North India. This chapter aims to explore the complex relationship between AARs and PM concentrations and to assess the associated risks of $PM_{2.5}$ in 526 cities worldwide. Understanding the aerosols and PM concentrations is beneficial for devising effective strategies to mitigate air pollution and safeguard public health globally.

4.2 Overview of Aerosol Optical Depth (AOD) and Particulate Matter (PM)

Aerosols encompass a mixture of suspended solid and liquid particles found in the atmosphere which includes dust, organic particles, industrial emissions, fog and mist, etc. (Kuniyal and Guleria 2019). However, the integral of the vertical aerosol extinction coefficient from the Earth's surface to the top of the atmosphere (TOA) is represented by AOD also called aerosol optical thickness (AOT) (Eq. 4.1) (Eddy 2012). The quantity of optical light extinction caused by aerosol scattering and absorption in the atmosphere is measured by AOD. Because they scatter and absorb solar and terrestrial radiation, these aerosol particles have a direct impact on the Earth's atmospheric energy balance (Zhou et al. 2023). Additionally, they indirectly change the physical, optical, and lifespan properties of clouds through cloud condensation nuclei (CCN) (Gallo et al. 2023). Equations 4.1 and 4.2 explain a strong correlation between AOD and factors such as particle size distribution, aerosol size, vertical profile, and radiation wavelength.

$$AOD/AOT = \int_{Surface}^{TOA} \sigma_{ext,z} dz \tag{4.1}$$

$$\sigma_{ext,z} = \pi \int Q_{ext}(m, r, \lambda) n(r) r^2 dr \tag{4.2}$$

where AOD/AOT is the aerosol optical depth/aerosol optical thickness, $\sigma_{ext,z}$ is the coefficient of aerosols at height z, r is the particle radius, λ is the wavelength, m is the refractive index, n(r) is the aerosol particle size distribution, and Q_{ext} is the function of m, r, and λ, respectively.

Cloud formation is significantly impacted by aerosol particles, which act as cloud condensation nuclei (CCNs) around which cloud droplets form (Gallo et al. 2023). The presence of a higher concentration of aerosols accelerates cloud formation compared to normal conditions (Buseck and Pósfai 1999). Aerosol particles can also change the properties of clouds that already exist, such as delaying the peak of precipitation during the day (Wang et al. 2020). Strong correlations have been found between satellite-based AOD and PM concentrations, especially when the majority of particles are localized in the boundary layer (Handschuh et al. 2022). Previous studies have also shown a linear correlation between satellite-based AOD and ground-level PM concentrations, although this relationship may vary due to differences in aerosol compositions, local climatic conditions, and geography, among other factors (Stirnberg et al. 2018; Su et al. 2018; Handschuh et al. 2022). Despite the correlation between satellite-based AOD and PM, several challenges exist in accurately estimating AOD and PM. These challenges include satellite sensor specifications, spatial resolution of satellite data, missing pixel values, surface reflectance, time separation among measurements, and the need for numerous auxiliary variables (Samuel et al. 2023). Enhancing the accuracy, consistency, and reliability of AOD and PM calculations, particularly regarding spatial distribution, should be the focus of efforts made to improve the performance of satellite data and related techniques (Zhu et al. 2023).

4.3 Remote Sensing Data for AOD Estimation

The availability of historical satellite remote sensing (RS) data has been helpful in analysing trends in AOD and PM concentrations within geographical regions. AOD levels and PM concentrations can be monitored using a variety of techniques that make use of different satellite datasets (Handschuh et al. 2022). Aerosol data is currently provided by multiple satellites and sensors, including the visible infrared imaging radiometer suite (VIIRS), ozone monitoring instrument (OMI), moderate resolution imaging spectroradiometer (MODIS), and advanced very high-resolution radiometer (AVHRR) (Ranjan et al. 2021). Furthermore, Light Detection and Ranging (LiDAR) serves as a highly efficient method for remote sensing of aerosols, featuring extensive detection capabilities, continuous monitoring, and superior spatial and temporal resolution (Ma et al. 2019). Table 4.1 provides an overview of operational satellite sensor products for obtaining aerosol data.

AOD levels across land and water have been estimated by numerous studies using AVHRR sensor data from National Oceanic and Atmospheric Administration (NOAA) series satellites. For example, Prasad and Gupta (1998) used AVHRR data at visible and near-infrared bands to estimate AOD levels over the Arabian Sea

Table 4.1 Satellite-based Aerosol Optical Depth (AOD) Products

AOD product	Satellite	Orbit	Wavelength	Data available since	Spatial–temporal coverage	Potential use
MODIS	Terra	Sun-synchronous	Visible, Infrared	1999	Global coverage, daily	Climate research, air quality monitoring
MISR	Terra	Sun-synchronous	Visible	1999	Global coverage, every 9 days	Aerosol characterization, cloud properties
MODIS	Aqua	Sun-synchronous	Visible, Infrared	2002	Global coverage, daily	Ocean colour observation, land surface monitoring
MAIAC	Terra & Aqua (MODIS)	Sun-synchronous	Visible	2000	Global coverage, daily	Land cover mapping, urbanization studies
OMI	Aura	Sun-synchronous	Visible, Ultraviolet	2003	Global coverage, daily	Ozone monitoring, air pollution detection
CALIOP	CALIPSO	Sun-synchronous	LiDAR (532 nm, 1064 nm)	2006	Global coverage, vertical profiles	Cloud and aerosol vertical profiling
GOCI	COMS	Geostationary	–	2010	East Asia region, hourly	Monitoring of East Asian aerosol transport
VIIRS	Suomi NPP	Sun-synchronous	Visible, Infrared	2011	Global coverage, daily	Wildfire detection, nighttime light observation
AHI	Himawari-8	Geostationary	–	2014	Asia–Pacific region, every 10 min	Weather forecasting, cloud tracking
ABI	GOES-16	Geostationary	–	2016	North and South America, every 5 min	Monitoring of atmospheric dynamics, severe weather detection

and compared the results with measurements made on the ground. AOD evaluation over Central Europe was studied by Hauser et al. (2005) utilizing AVHRR datasets and a bidirectional reflectance distribution function (BRDF) model for estimation. Similarly, Mei et al. (2014) established an empirical linear relationship between AOD levels and MODIS-based reflectance values to show the potential of AVHRR data in AOD estimation over the northeast region of China (Eq. 4.3).

$$R_{Aerosol}(\lambda, \mu_0, \mu, \varphi) = R_{TOA}(\lambda, \mu_0, \mu, \varphi) - R_{Ray}(\lambda, \mu_0, \mu, \varphi)$$
$$- \frac{A_{sfc}(\lambda) \times T_1(\lambda, \mu_0) \times T_2(\lambda, \mu)}{1 - A_{sfc}(\lambda) \times s(\lambda)} \quad (4.3)$$

where μ, μ_0, and φ are the solar zenith, satellite zenith, and relative azimuth angle, λ is the wavelength, $R_{Aerosol}(\lambda, \mu_0, \mu, \varphi)$ is the aerosol reflectance, $R_{TOA}(\lambda, \mu_0, \mu, \varphi)$ is top of the atmosphere reflectance, $R_{Ray}(\lambda, \mu_0, \mu, \varphi)$ is the Rayleigh reflectance, $A_{sfc}(\lambda)$ and $s(\lambda)$ is the surface spectral albedo and atmospheric hemispherical albedo, $T_1(\lambda, \mu_0)$ and $T_2(\lambda, \mu)$ is the total transmission of light propagating downward and upward from the surface to the top of the atmosphere, respectively.

In recent years, Hsu et al. (2017) introduced a novel method for assessing the nature of aerosols over land and ocean using AVHRR datasets, extending existing algorithms. The normalized difference vegetation index (NDVI) and minimal reflectance framework were used in this method to lessen the impact of shifting vegetation on surface reflectance. Furthermore, MODIS sensors onboard Terra and Aqua satellites have revolutionized AOD and PM estimation. With comprehensive spectral coverage and improved spatial resolution, MODIS-based aerosol products offer enhanced capabilities for AOD estimation compared to AVHRR data. The deep blue (DB) and dark target (DT) algorithms designed for MODIS data enable AOD estimation over various surface types, addressing limitations encountered with AVHRR datasets. For global aerosol monitoring, the MODIS-based AOD products with DT and DB algorithms are readily accessible. These developments in remote sensing technology highlight how crucial it is to use satellite data for precise and trustworthy AOD and PM concentration estimation, supporting extensive studies on air quality and environmental monitoring initiatives.

4.4   Satellite-Based Particulate Matter (PM) Concentration Estimation

Various methodologies, including empirical/semi-empirical models, statistical models, vertical correlation models, and chemical transport models, have been employed to estimate PM concentrations using satellite-based AOD datasets, alongside supporting variables (Fig. 4.1). Sections 4.2 and 4.3 primarily focus on reviewing the satellite-based data, calculation of AOD for the computation of PM. However, this section underscores the possibility of PM computation through empirical models based on the primary aerosols. These models are valued for their ease of use, quickness, applicability, and superiority over alternative techniques for predicting PM concentrations at the ground level using satellite-based datasets.

Despite the availability of numerous approaches for PM estimation from satellite data, accurately estimating PM remains a complex issue due to various influencing factors, including land-use types, meteorological parameters, and cloud contamination, among others. Consequently, the relationship between aerosol properties and PM is typically evaluated by considering local conditions. Challenges such as the coarser spatial resolution of satellite data and missing pixel values pose significant obstacles to precise ground-level PM estimation from satellite data. Chu et al. (2016) have provided a comprehensive review of the capabilities and limitations of various satellite remote sensing approaches for ground-level PM concentration estimation. Furthermore, this chapter also evaluates the PM concentrations from the satellite-based products in Sect. 4.4.1 and the associated risk of PM globally.

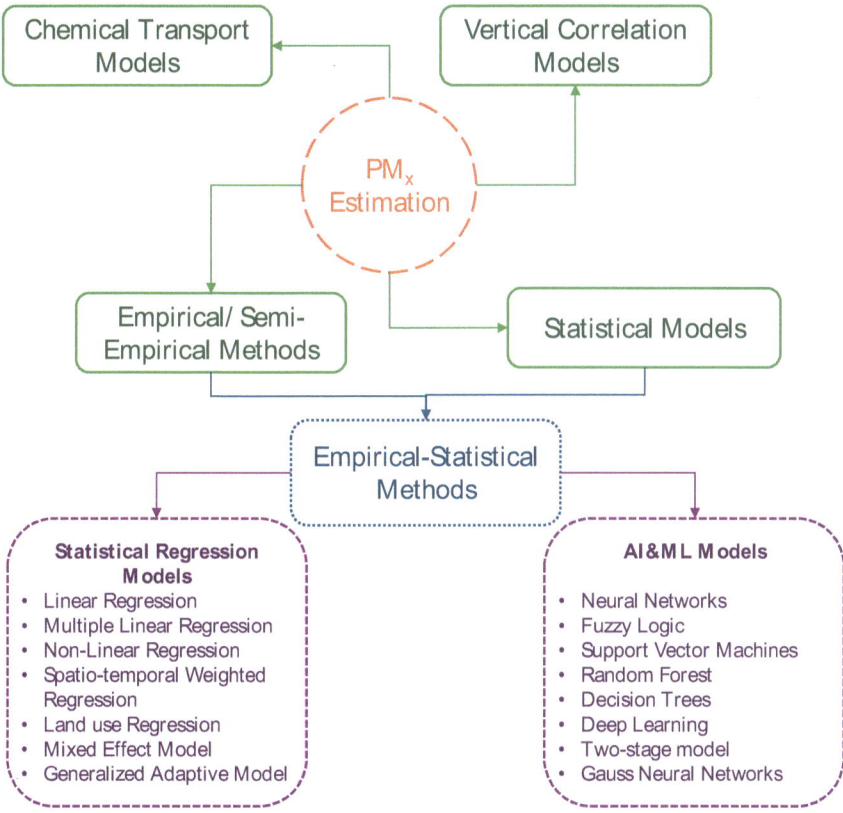

Fig. 4.1 Commonly used methods for the estimation of PM_x (where x denotes the particle size in micrometres) based on AOD/AOT satellite products

4.5   Estimation of Particulate Matter Concentrations from Satellite Reanalysis Datasets

4.5.1   Datasets

This study uses datasets with a spatio-temporal resolution of $0.5° \times 0.625°$ and 1 h, spanning the period from January 1, 1980, to December 31, 2023, using MERRA-2 reanalysis data that was obtained from the NASA GESDISC DATA ARCHIVE application (Randles et al. 2017). The dataset comprises 53 different atmospheric aerosol-related variables, such as time, latitude, and longitude. Furthermore, using the climate data operator (CDO), the following data is extracted: time (t), latitude (ϕ), longitude (λ), black carbon surface mass concentration (BC_{SMASS}), dust surface mass concentration-$PM_{2.5}$ ($DU_{SMASS25}$), organic carbon surface mass concentration (OC_{SMASS}), sea salt surface mass concentration-$PM_{2.5}$ ($SS_{SMASS25}$), and sulphate surface mass concentration (SO_{4SMASS}). In addition, the hourly temporal resolution is translated into the daily by taking the mean of the next 24 h. According to Buchard et al. (2016) and Provençal et al. (2017), the $PM_{2.5}$ concentration ($\mu g/m^3$) for every grid cell ($0.5° \times 0.625°$) is calculated as (Eq. 4.4).

$$PM_{2.5} = (BC_{SMASS} + DU_{SMASS25} + OC_{SMASS} + SS_{SMASS25} + 1.375 \times SO_{4SMASS}) \times 10^9 \quad (4.4)$$

4.5.2 $PM_{2.5}$ Concentration and Trends

From 1980 to 2023, this study examined the correlation between MERRA-2 satellite reanalysis datasets and $PM_{2.5}$ concentration data collected from 526 cities worldwide. Results showed a strong agreement between the reanalysis and measured datasets, with a Root Mean Square Error (RMSE) averaging 7.34 $\mu g/m^3$ (Fig. 4.2a), indicating minimal discrepancy. Taylor's diagram revealed a high correlation coefficient of 0.87 (Fig. 4.2a), suggesting a linear relationship between the datasets. The coefficient of determination (R2) was calculated at 0.75, indicating that 75% of the variability in measured data can be explained by the reanalysis datasets (Fig. 4.2c). These findings demonstrate robust agreement between MERRA-2 reanalysis datasets and measured air quality across studied cities. $PM_{2.5}$ concentration trends varied across regions from 1980 to 2023, with the Indo-Gangetic Plains experiencing a notable increase of 1–1.5 $\mu g/m3$ (Fig. 4.2b). The Taklamakan Desert region and Eastern China/western Africa saw smaller rises of 0.5–1 $\mu g/m3$ and 0–0.5 $\mu g/m3$, respectively. Certain regions, including Niger, Chad, Gulf countries, and India, showed consistently high $PM_{2.5}$ levels with increasing trends, highlighting a significant air pollution problem. Conversely, the European Union and Eastern USA displayed lower $PM_{2.5}$ concentrations with decreasing trends, indicating progress towards improved air quality.

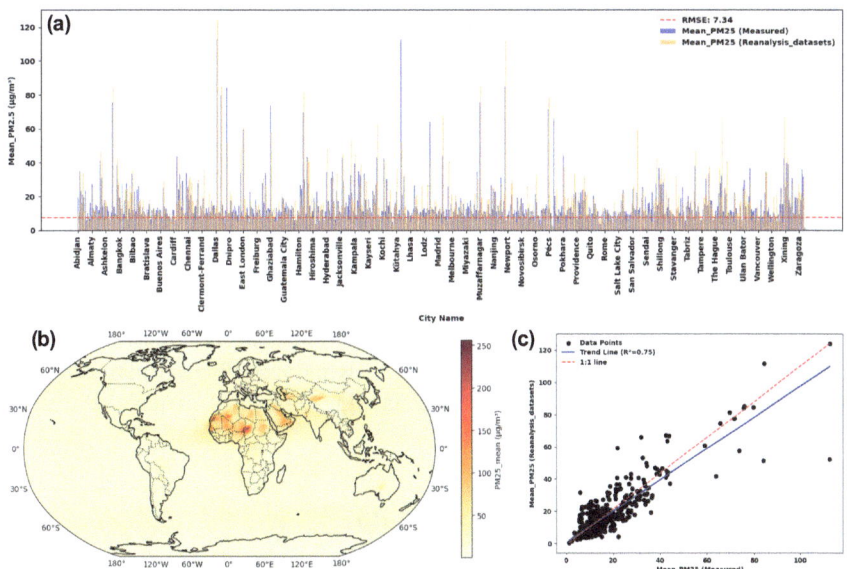

Fig. 4.2   a Observed and estimated PM$_{2.5}$ over 526 cities, **b** Mean PM$_{2.5}$ concentrations, and **c** Correlation between observed and estimated PM$_{2.5}$ during the period 1980–2023

These trends underscore the impact of human activities on PM$_{2.5}$ concentrations, with implications for society and ecosystems.

4.6   Aerosol Atmospheric Rivers and Particulate Matter (PM)

4.6.1   Correlation and Spatial Selectivity

The phenomenon of AARs and its correlation with PM concentrations reveals crucial insights into the dynamics of aerosols and their impact on air quality. AARs, composed of key aerosol species such as black carbon, dust, organic carbon, sea salt, and sulfates, play a pivotal role in increasing PM concentrations within the regions they cover (Chakraborty et al. 2021). These aerosol species constitute primary components of PM, thereby significantly influencing its concentration levels. In the context of aerosol transport, AARs emerge as potent contributors, responsible for up to 80% of the observed increase in PM concentration within their respective regions (Chakraborty et al. 2022). This underscores the substantial influence AARs exert on local air quality dynamics. The presence of AARs leads to a notable surge in PM levels, exacerbating air pollution concerns and posing significant challenges for environmental and public health management. Figure 4.3 serves as compelling evidence of the correlation between AARs and PM concentrations. The visual representation depicted in the figure elucidates the intricate relationship between these two phenomena, highlighting their co-occurrence and mutual influence. Through empirical observation and data analysis, the figure underscores the interconnectedness of AARs and PM concentrations, reaffirming their associated relationship. However, it's essential to note certain characteristics of AARs that shape their impact on PM concentrations. AARs, often spanning lengths exceeding 2000 km, exhibit varying spatial extents, influencing the distribution and magnitude of PM concentrations across different regions. While AARs significantly contribute to elevated PM levels within their coverage areas, their extensive reach also implies the exclusion of smaller regions characterized by localized high PM concentrations. This spatial selectivity underscores the nuanced nature of AAR-induced PM transport and distribution, necessitating a comprehensive understanding of regional air quality dynamics (Fig. 4.3).

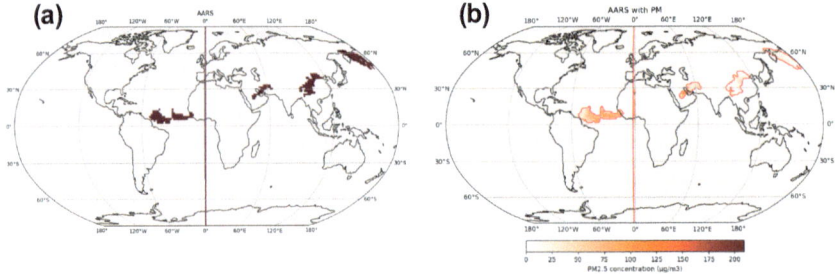

Fig. 4.3 **a** AARs and **b** corresponding PM$_{2.5}$ concentration for the particular AAR

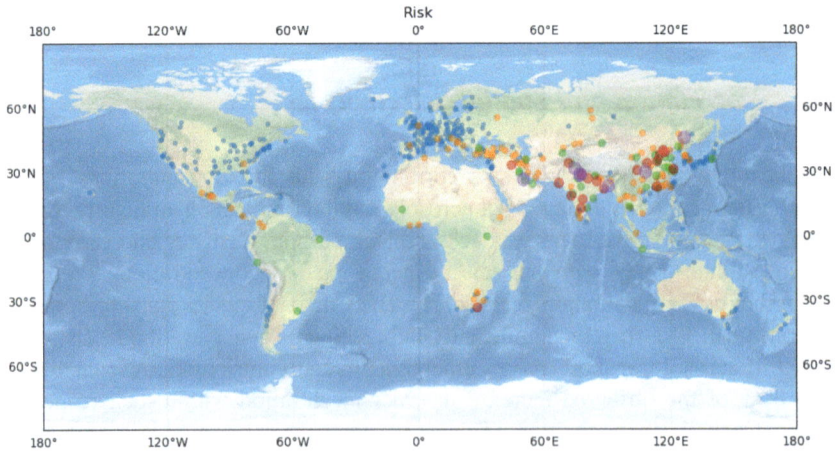

Fig. 4.4 Computed PM$_{2.5}$ risk among 526 cites over the globe

4.6.2 PM₂.₅ Risk Assessment

The chapter also evaluates the risk posed by $PM_{2.5}$ through aerosol pollution across various cities using a dual approach: hazard-based risk assessment and vulnerability analysis incorporating population data from 526 cities. Cities are categorized into five tiers based on risk: very low, low, medium, high, and very high. Notably, Lahore, Dhaka, Dammam, and New Delhi, are identified as highly vulnerable, indicating significant risk. Kolkata, Ghaziabad, Meerut, and Muzaffarnagar and Baghdad follow closely, classified as high-risk areas (Fig. 4.4). Additionally, 25 cities fall into the medium-risk category, while 85 cities are deemed low risk, with a vast majority facing very low-risk levels. Zurich emerges as a cleanliness example among the cities studied. However, when factoring population density, New Delhi, Lahore, and Dhaka stand out as highly susceptible to extreme pollution due to dense populations, with six other cities also identified as high-risk areas.

Further analysis reveals that cities with lower populations, like Perth, Richards Bay, and Hobart, exhibit less vulnerability to extreme pollution. Conversely, cities in low-income brackets or categorized as underdeveloped often fall within the medium-risk zone, especially when population density is not considered. Moreover, regions within the Indo-Gangetic Plain (IGP) exhibit varying risk levels, influenced by pollution levels and socioeconomic conditions.

The study introduces a novel categorization of risk into six tiers—High Safety, Safety, Deep Stabilization, Risk, Stabilization, and High Risk—based on observed trends and average annual $PM_{2.5}$ concentrations globally. Within the European Union (EU), countries are classified under High Safety due to declining trends in $PM_{2.5}$ levels and concentrations below 30 $\mu g/m^3$. Conversely, the USA and Canada are designated as Safety zones despite upward trends, given their $PM_{2.5}$ concentrations below the threshold. However, historical trends suggest a potential escalation in $PM_{2.5}$ concentrations in these regions. In South America, countries like Argentina and Paraguay are categorized as Deep Stabilization, reflecting decreasing trends and concentrations between 30–60 $\mu g/m^3$. Conversely, Russia and western China are at risk due to significant concentrations and increasing trends in $PM_{2.5}$ levels. Areas near the western Sahara and Indonesia are undergoing Stabilization despite high concentrations, requiring concerted efforts to stabilize. The most severe classification, High Risk, applies to regions in eastern China, India, and Niger, grappling with very high concentrations and upward trajectories in $PM_{2.5}$ concentrations, necessitating urgent interventions to mitigate air pollution risks.

4.7 Conclusion

This chapter provides a comprehensive exploration of the dynamics of air pollution, particularly focusing on the estimation of PM concentrations through satellite reanalysis datasets and the assessment of associated risks globally. The escalating issue of air pollution has gained significant attention in recent decades, with detrimental impacts on human health, the environment, and the economy. The complex relationship between aerosol optical depth (AOD), PM concentrations, and their monitoring through satellite remote sensing technologies underscores the importance of accurate and reliable estimation methods, given the profound implications of air pollution on public health and climate.

Satellite reanalysis datasets, notably MERRA-2, offer valuable insights into PM concentrations, enabling researchers to assess long-term trends and variations across different regions. The study's findings reveal a robust agreement between MERRA-2 reanalysis datasets and measured $PM_{2.5}$ concentrations, indicating the effectiveness of satellite remote sensing in monitoring air quality. Moreover, the analysis of $PM_{2.5}$ concentration trends highlights the complex interplay of various factors, including regional pollution sources, meteorological conditions, and socioeconomic factors.

Aerosol Atmospheric Rivers (AARs) emerge as significant contributors to PM concentrations, with their influence extending across large geographic areas. The correlation between AARs and PM concentrations underscores the need for a comprehensive understanding of aerosol transport dynamics to address air quality challenges effectively. Furthermore, the risk assessment of $PM_{2.5}$ pollution in cities worldwide reveals varying levels of vulnerability, with densely populated urban centres facing heightened risks. The categorization of regions based on observed $PM_{2.5}$ trends and concentrations provides valuable insights for policymakers and stakeholders to prioritize interventions and allocate resources effectively.

References

Anggraini TS, Irie H, Sakti AD, Wikantika K (2024) Machine learning-based global air quality index development using remote sensing and ground-based stations. Environ Adv 15:100456. https://doi.org/10.1016/j.envadv.2023.100456

Anwar MN, Shabbir M, Tahir E et al (2021) Emerging challenges of air pollution and particulate matter in China, India, and Pakistan and mitigating solutions. J Hazard Mater 416:125851. https://doi.org/10.1016/j.jhazmat.2021.125851

Balakrishnan K, Dey S, Gupta T et al (2019) The impact of air pollution on deaths, disease burden, and life expectancy across the states of India: the Global Burden of Disease Study 2017. Lancet Planet Heal 3:e26–e39. https://doi.org/10.1016/S2542-5196(18)30261-4

Buchard V, da Silva AM, Randles CA et al (2016) Evaluation of the surface PM2.5 in Version 1 of the NASA MERRA Aerosol Reanalysis over the United States. Atmos Environ 125:100–111. https://doi.org/10.1016/j.atmosenv.2015.11.004

Buseck PR, Pósfai M (1999) Airborne minerals and related aerosol particles: Effects on climate and the environment. Proc Natl Acad Sci 96:3372–3379. https://doi.org/10.1073/pnas.96.7.3372

Chakraborty S, Guan B, Waliser DE, et al (2021) Extending the atmospheric river concept to aerosols: climate and air quality impacts. Geophys Res Lett 48.https://doi.org/10.1029/2020GL091827

Chakraborty S, Guan B, Waliser DE, da Silva AM (2022) Aerosol atmospheric rivers: climatology, event characteristics, and detection algorithm sensitivities. Atmos Chem Phys 22:8175–8195. https://doi.org/10.5194/acp-22-8175-2022

Chand K, Kuniyal JC, Kanga S et al (2021) Aerosol characteristics and their impact on the Himalayan energy budget. Sustainability 14:179. https://doi.org/10.3390/su14010179

Eddy C (2012) Aerosol direct radiative forcing: a review. In: Atmospheric aerosols—regional characteristics—chemistry and physics. InTech

Fuller R, Landrigan PJ, Balakrishnan K et al (2022) Pollution and health: a progress update. Lancet Planet Heal 6:e535–e547. https://doi.org/10.1016/S2542-5196(22)00090-0

Gallo F, Uin J, Sanchez KJ et al (2023) Long-range transported continental aerosol in the eastern North Atlantic: three multiday event regimes influence cloud condensation nuclei. Atmos Chem Phys 23:4221–4246. https://doi.org/10.5194/acp-23-4221-2023

Guleria RP, Chand K (2020) Emerging patterns in global and regional aerosol characteristics: a study based on satellite remote sensors. J Atmos Solar-Terrestrial Phys 197:105177. https://doi.org/10.1016/j.jastp.2019.105177

Handschuh J, Erbertseder T, Schaap M, Baier F (2022) Estimating PM2.5 surface concentrations from AOD: A combination of SLSTR and MODIS. Remote Sens Appl Soc Environ 26:100716. https://doi.org/10.1016/j.rsase.2022.100716

Hauser A, Oesch D, Foppa N, Wunderle S (2005) NOAA AVHRR derived aerosol optical depth over land. J Geophys Res Atmos 110.https://doi.org/10.1029/2004JD005439

Hsu NC, Lee J, Sayer AM et al (2017) Retrieving near-global aerosol loading over land and ocean from AVHRR. J Geophys Res Atmos 122:9968–9989. https://doi.org/10.1002/2017JD026932

Jacobs ET, Burgess JL, Abbott MB (2018) The Donora smog revisited: 70 years after the event that inspired the clean air act. Am J Public Health 108:S85–S88. https://doi.org/10.2105/AJPH.2017.304219

Krishan M, Jha S, Das J et al (2019) Air quality modelling using long short-term memory (LSTM) over NCT-Delhi, India. Air Qual Atmos Heal 12:899–908. https://doi.org/10.1007/s11869-019-00696-7

Kuniyal JC, Guleria RP (2019) The current state of aerosol-radiation interactions: a mini review. J Aerosol Sci 130:45–54. https://doi.org/10.1016/j.jaerosci.2018.12.010

Li C, van Donkelaar A, Hammer MS et al (2023) Reversal of trends in global fine particulate matter air pollution. Nat Commun 14:5349. https://doi.org/10.1038/s41467-023-41086-z

Ma X, Wang C, Han G et al (2019) Regional atmospheric aerosol pollution detection based on LiDAR remote sensing. Remote Sens 11:2339. https://doi.org/10.3390/rs11202339

Mei LL, Xue Y, Kokhanovsky AA et al (2014) Retrieval of aerosol optical depth over land surfaces from AVHRR data. Atmos Meas Tech 7:2411–2420. https://doi.org/10.5194/amt-7-2411-2014

Prasad S, Gupta R (1998) Estimation and evaluation of Aerosol Optical Depth using NOAA AVHRR data. Adv Sp Res 22:1525–1528. https://doi.org/10.1016/S0273-1177(99)00020-4

Provençal S, Buchard V, da Silva AM et al (2017) Evaluation of PM surface concentrations simulated by Version 1 of NASA's MERRA Aerosol Reanalysis over Europe. Atmos Pollut Res 8:374–382. https://doi.org/10.1016/j.apr.2016.10.009

Randles CA, da Silva AM, Buchard V et al (2017) The MERRA-2 Aerosol Reanalysis, 1980 onward. Part I: system description and data assimilation evaluation. J Clim 30:6823–6850. https://doi.org/10.1175/JCLI-D-16-0609.1

Ranjan AK, Patra AK, Gorai AK (2021) A review on estimation of particulate matter from satellite-based aerosol optical depth: data, methods, and challenges. Asia-Pacific J Atmos Sci 57:679–699. https://doi.org/10.1007/s13143-020-00215-0

Rautela KS, Singh S, Goyal MK (2024a) Characterizing the spatio-temporal distribution, detection, and prediction of aerosol atmospheric rivers on a global scale. J Environ Manage 351:119675. https://doi.org/10.1016/j.jenvman.2023.119675

Rautela KS, Singh S, Goyal MK (2024b) Resilience to air pollution: A novel approach for detecting and predicting aerosol atmospheric rivers within earth system boundaries earth systems and environment. https://doi.org/10.1007/s41748-024-00421-0

Samuel C, Kumar RP, Gautam S (2023) Review of retrieval of aerosol optical depth to estimate particle concentration and its challenges based on spatiotemporal relationships by various spectroradiometer models. Geol J 58:4403–4416. https://doi.org/10.1002/gj.4780

Shakya D, Deshpande V, Goyal MK, Agarwal M (2023) PM2.5 air pollution prediction through deep learning using meteorological, vehicular, and emission data: a case study of New Delhi, India. J Clean Prod 427:139278. https://doi.org/10.1016/j.jclepro.2023.139278

Stirnberg R, Cermak J, Andersen H (2018) An analysis of factors influencing the relationship between satellite-derived AOD and ground-level PM10. Remote Sens 10. https://doi.org/10.3390/rs10091353

Su T, Li Z, Kahn R (2018) Relationships between the planetary boundary layer height and surface pollutants derived from lidar observations over China: regional pattern and influencing factors. Atmos Chem Phys 18:15921–15935. https://doi.org/10.5194/acp-18-15921-2018

Wang Y, Zheng X, Dong X et al (2020) Impacts of long-range transport of aerosols on marine-boundary-layer clouds in the eastern North Atlantic. Atmos Chem Phys 20:14741–14755. https://doi.org/10.5194/acp-20-14741-2020

WHO (2021) WHO global air quality guidelines: particulate matter (PM2.5 and PM10), ozone, nitrogen dioxide, sulfur dioxide and carbon monoxide. World Health Organization

Xian P, Reid JS, Atwood SA et al (2013) Smoke aerosol transport patterns over the Maritime Continent. Atmos Res 122:469–485. https://doi.org/10.1016/J.ATMOSRES.2012.05.006

Zhou C, Duan Z, Ling X, Yang Y (2023) Effects of local aerosol and transported dust pollution on the surface energy balance over farmland in eastern China during spring. Front Environ Sci 10. https://doi.org/10.3389/fenvs.2022.1059292

Zhu S, Tang J, Zhou X et al (2023) Research progress, challenges, and prospects of PM 2.5 concentration estimation using satellite data. Environ Rev 31:605–631. https://doi.org/10.1139/er-2022-0125

Chapter 5
Mitigation, Adaptation, and Resilience Strategies for Extreme Aerosol Pollution

Abstract Extreme aerosol pollution through aerosol atmospheric rivers (AARs) presents significant challenges to public health and environmental quality, particularly in urban areas where concentrations are often highest. This chapter explores adaptation and resilience strategies aimed at mitigating the impacts of AARs. Urban planning and design interventions, including the integration of green infrastructure and promotion of sustainable development practices, play a crucial role in minimizing aerosol exposure. Green infrastructure initiatives, such as increasing urban greenery and implementing green roofs and permeable pavements, offer promising avenues for reducing pollutant concentrations and mitigating heat-related health risks. Improved ventilation and filtration systems in buildings contribute to maintaining indoor air quality and promoting occupant health and well-being. Health and education programmes targeting vulnerable populations raise awareness about the health risks associated with aerosol pollution and advocate for policies to improve air quality. Early warning systems, leveraging remote sensing and monitoring technologies, provide critical alerts for extreme aerosol events, enabling timely evacuation and emergency preparedness. Integration of aerosol data into climate models enhances forecasting accuracy and informs adaptive measures to mitigate the impacts of aerosol pollution on weather patterns. Community-based resilience initiatives empower local communities to address air quality issues through participatory monitoring programmes, educational campaigns, and policy development efforts, fostering a sense of ownership and empowerment in the fight against aerosol pollution. Overall, these adaptation and resilience strategies are essential for enhancing societal preparedness and response capabilities to mitigate the impacts of aerosol pollution events and safeguard public health and environmental quality.

Keywords Adaptation strategies · Aerosol pollution · Early warning systems · Resilience initiatives · Urban planning

5.1 Introduction

Extreme aerosol pollution refers to a severe concentration of aerosol particles in the atmosphere, leading to significant health risks and environmental impacts. Aerosol atmospheric rivers (AARs) contribute to severe air pollution, posing threats that require urgent measures to mitigate their harmful effects (Rautela et al. 2024a). AARs affect many aspects of the natural environment, including biodiversity loss, disturbances to ecosystems, and changes in atmospheric composition (Haldane et al. 2019; Rautela et al. 2024b). Aerosols, which are made up of a wide variety of particles suspended in the atmosphere, have a major impact on Earth's radiative balance and hence play a role in the dynamics of climate change. These tiny particles can either scatter or absorb solar energy, changing the climate system locally by warming or cooling (Barthlott et al. 2022). Furthermore, changes in cloud characteristics and precipitation patterns brought on by aerosols amplify climatic variability and raise the frequency of extreme weather events worldwide (Dubey and Goyal 2020; Liu et al. 2020; Kumar et al. 2021).

Air pollution through AARs has an equally severe negative impact on human health, with respiratory and cardiovascular disorders emerging as the main causes of concern (Fuller et al. 2022). These AARs are the prime constituents of fine particulate matter ($PM_{2.5}$) that can enter the lungs deeply through inhalation and present a serious risk to respiratory health (Chakraborty et al. 2021). This may result in increased vulnerability to respiratory tract infections, oxidative stress, and inflammation. Long-term exposure to high $PM_{2.5}$ levels has been repeatedly associated in studies with a higher risk of lung cancer, asthma, chronic obstructive pulmonary disease (COPD), and cardiovascular death (Jiang et al. 2016). Children, the elderly, and people with pre-existing medical issues are among the vulnerable groups that are more susceptible to the negative impacts of aerosol exposure, which exacerbates already-existing health inequities (Park et al. 2020). The fine particulate matter ($PM_{2.5}$) can enter the lungs deeply through inhalation, it is a common aerosol component that presents a serious risk to respiratory health (Park et al. 2020).

Adaptation strategies complement mitigation efforts by focusing on minimizing exposure and vulnerability to aerosol pollution (Kaur and Pandey 2021). Urban planning initiatives, such as the development of green spaces and pedestrian-friendly infrastructure, can help mitigate the impacts of aerosol pollution in densely populated areas (Barwise and Kumar 2020). Moreover, public health interventions, including the establishment of air quality monitoring networks and the provision of respiratory health services, are critical for protecting vulnerable populations from the adverse effects of aerosol exposure (Kelly and Fussell 2015). In Beijing, China, the adoption of measures such as vehicle emission controls and the closure of coal-fired power plants has led to a significant reduction in $PM_{2.5}$ concentrations and an improvement in air quality over the past decade (Lu et al. 2020).

Initiatives aimed at strengthening societal resilience are essential for improving readiness and response capacities in the case of aerosol pollution (Sturiale 2019). Early warning systems, supported by advancements in meteorological forecasting and remote sensing technology, enable timely evacuation and emergency planning in the case of severe aerosol pollution events through AARs (Mo et al. 2019). Initiatives for community-based resilience that are fueled by grassroots mobilization and local knowledge promote social cohesion and flexible governance frameworks, which make it easier to respond to extreme aerosol pollution through AARs incidents in a coordinated manner (Shammin et al. 2022).

5.2 Mitigation Strategies

Reducing anthropogenic emissions, creating cutting-edge technology, and improving natural removal mechanisms are all necessary for mitigating excessive aerosol pollution (Fig. 5.1). The ways to reduce aerosol emissions and lessen their negative effects on the environment and human health are examined in this section.

5.2.1 Reduction of Anthropogenic Aerosol Emissions

5.2.1.1 Policies and Regulations Targeting Key Sources

Establishing emission standards, advancing cleaner technology, and providing incentives for pollution reduction strategies are all important ways that policies and regulations help to mitigate aerosol pollution. Around the world, governments have put in place several regulations aimed at major sources of aerosol emissions, such as automobile exhaust and industrial facilities. However, various regulations pertaining to industrial emissions, such as emissions trading programmes, emission standards, and pollution levies, are designed to minimize the atmospheric release of precursor gases

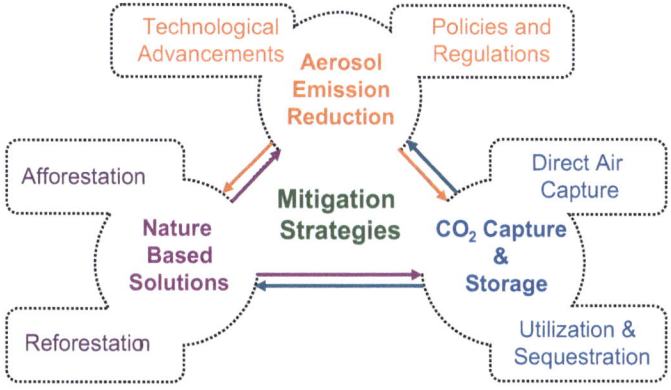

Fig. 5.1 Possible Mitigation Strategies for the reduction of aerosol pollution

and aerosols. For example, manufacturing facilities, refineries, and power plants are subject to strict emission restrictions set by the European Union's Industrial Emissions Directive. Comparably, the United States Clean Air Act requires the Environmental Protection Agency (EPA) to control air pollutants from industrial sources, including aerosols.

Aerosol emissions from exhaust fumes have been largely mitigated in the transportation sector by strict vehicle emissions rules, fuel economy laws, and incentives for electric vehicles. Vehicle emission regulations similar to those of Euro 6 have been imposed in nations like China and India, necessitating the use of low-sulfur fuels and the development of cleaner engine technologies. In addition, policies like congestion pricing and vehicle scrappage programmes try to lower emissions and traffic jams in urban areas.

5.2.1.2 Technological Advancements for Cleaner Production Processes

Through the development of cleaner production processes, increased energy efficiency, and reduced emissions from industrial activities, technological advancements play a critical role in mitigating aerosol pollution. Aerosols and polluting gases can now be eliminated from industrial exhaust streams because of developments in pollution control technology like electrostatic precipitators, scrubbers, and catalytic converters as discussed in Chap. 1.

Furthermore, by reducing aerosol emissions, research and development initiatives concentrate on creating sustainable industrial techniques, renewable energy sources, and substitute fuels. Sulfur dioxide (SO_2) emissions, a precursor to sulfate aerosols, can be greatly reduced, for example, by switching from coal-fired power plants to greener energy sources like natural gas, wind, and solar power. Furthermore, automobiles can now achieve improved fuel efficiency and lower emissions of aerosols and nitrogen oxides (NOx) because of developments in engine technology, such as fuel injection systems, exhaust gas recirculation (EGR), and selective catalytic reduction (SCR). Since they emit no tailpipe emissions and help to clean up the air in metropolitan areas, electric vehicles (EVs) provide a viable way to reduce aerosol pollution.

5.2.2 Carbon Capture and Storage (CCS) Technologies

5.2.2.1 Direct Air Capture (DAC) of Aerosols

Carbon capture and storage (CCS) technologies offer a promising approach to mitigate aerosol pollution by capturing aerosols and greenhouse gases directly from the atmosphere and storing them underground. Despite significant efforts to reduce CO_2 emissions, the annual release still amounts to tens of billions of tonnes, prompting the IPCC to advocate for CO_2 capture technology to curb the rise in atmospheric

CO_2 levels (Wang et al. 2021). Traditionally, CO_2 capture has focused on mitigating emissions from major stationary sources like power plants, cement factories, oil refineries, and metal production facilities (Bandilla 2020). Mostly, direct air capture (DAC) technologies utilize chemical sorbents or absorbents to capture CO_2, from ambient air. Direct air capture (DAC) of CO_2, first conceptualized in 1999 by Lackner, has evolved into a rapidly growing environmental technology with a shift towards experimental work and increased publications in the past decade (Sanz-Pérez et al. 2016). DAC offers advantages over conventional CO_2 capture by addressing emissions from distributed sources and being location-flexible, without the challenges posed by contaminants in flue gas (Sanz-Pérez et al. 2016; García-Bordejé and González-Olmos 2024). While uncertainties persist regarding the long-term storage and sequestration of captured CO_2, DAC presents opportunities for utilization as a chemical feedstock, fuel synthesis, or other value-added products (Bajpai et al. 2022). Sorbent-based processes, particularly those employing chemisorbent materials, have shown promise in capturing CO_2 efficiently from ambient air, highlighting DAC's potential in climate change mitigation and carbon utilization efforts (Priyadarshini et al. 2023).

However, DAC technology is effective in capturing carbon emissions. Similarly, this technology could be used to selectively capture fine particulate matter (PM) and aerosol particles from the ambient air in industries such as thermal power plants, cement factories, oil refineries, etc. Once captured, the aerosols can be separated from the sorbent material using heat or pressure, allowing for their utilization or sequestration. DAC technologies have the potential to remove significant quantities of aerosol gases from the atmosphere, thereby mitigating their environmental impacts.

5.2.2.2 Utilization and Sequestration of Captured Aerosols

Captured aerosols can be utilized or sequestered to prevent their release into the atmosphere and mitigate their environmental impacts. CO_2 utilization, encompassing both carbon capture and utilization (CCU) and carbon dioxide removal (CDR) technologies, represents a multifaceted approach to combating climate change while fostering economic growth (Nagireddi et al. 2024). CCU attempts to transform captured CO_2 into valuable products, including chemicals, fuels, and building materials, thereby reducing greenhouse gas emissions, and promoting circular economy principles (D'Alessandro et al. 2010). Methods like chemical, biological, and electrochemical conversion facilitate this process, offering alternatives to fossil-based materials and reducing reliance on finite resources (Sneddon et al. 2014; Hu et al. 2015; Zhao et al. 2016; Kätelhön et al. 2019; Jo et al. 2020). Concurrently, CDR technologies aim to remove CO_2 from the atmosphere, utilizing methods such as afforestation, direct air capture, and mineral carbonation (Nagireddi et al. 2024). Despite the promise of CO_2 utilization, challenges persist, including energy requirements, scalability issues, and concerns over environmental impacts (Li et al. 2022). However, collaboration across industries offers opportunities for innovation, paving the way for large-scale deployment of CO_2 utilization technologies and the creation of a more sustainable future.

When it comes to long-term carbon capture and storage (CCS), geological storage is a top option in the field of CO_2 sequestration (Ajayi et al. 2019). This method involves injecting captured CO_2 into geological formations, such as depleted oil and gas reservoirs or saline formations, effectively trapping the greenhouse gas underground for thousands of years (Ketzer et al. 2012). Geological sequestration holds immense potential for reducing GHG emissions and mitigating climate change, with large-scale demonstration projects showcasing its feasibility (Shaw and Mukherjee 2022). However, challenges, including high costs and environmental risks, underscore the need for rigorous monitoring and management protocols. Meanwhile, alternative methods like ocean storage and mineralization offer additional avenues for CO_2 sequestration, albeit with their own sets of technical and environmental problems (Nagireddi et al. 2024). Despite these challenges, ongoing research and innovation offer hope for overcoming barriers and unlocking the full potential of CO_2 sequestration as a crucial tool in the fight against climate change.

Utilization and sequestration technologies could offer indirect benefits for aerosol removal from the atmosphere by addressing the root cause of climate change. Hazardous atmospheric aerosols, which are also a major constituent of particulate matter from industrial emissions or wildfires, contribute to air pollution and pose significant health risks (Southerland et al. 2022). By reducing emissions through developing similar technologies such as carbon capture and utilization (CCU) and carbon dioxide removal (CDR) technologies, these methods indirectly mitigate climate change, which can lead to a more stable climate. A stable climate reduces the frequency and intensity of events like wildfires and industrial pollution, thereby limiting the release of hazardous aerosols into the atmosphere. Additionally, geological storage of captured atmospheric aerosols could promote cleaner energy sources and reduce the reliance on fossil fuels, significant contributors to aerosol emissions. Thus, while not directly targeting aerosol removal, CO_2 utilization and sequestration technologies play a crucial role in mitigating the factors that contribute to hazardous atmospheric aerosols, thereby promoting cleaner air and improving public health.

5.2.3 Nature-Based Solutions to Enhance Natural Aerosol Removal Processes

Nature-based solutions such as afforestation and reforestation mitigate aerosol pollution by enhancing aerosol removal processes through vegetation uptake and deposition (Cunningham et al. 2015). Aerosols and other airborne pollutants are drawn into plant tissues and soil by trees and other vegetation, which act as natural filters (Barwise and Kumar 2020). The aim of reforestation projects is to increase the amount of vegetation cover and restore ecosystem services by planting trees in degraded or barren areas (Kim et al. 2021). Afforestation and reforestation efforts have the potential to enhance natural aerosol removal processes through the expansion and restoration of forested areas (Hasegawa et al. 2024). Research findings suggest that carefully selecting carbon-intensive forest types for afforestation could significantly increase carbon sequestration rates, thereby indirectly benefiting aerosol removal (Gao and Li 2023). By maximizing carbon sequestration potential, afforestation measures could contribute to a 25% increase in carbon sequestration compared to native forest types (Hasegawa et al. 2024). Furthermore, when combined with complementary food measures, this potential could be further elevated, reaching up to a 49% increase in carbon sequestration compared to native forests (Hasegawa et al. 2024). These findings underscore the importance of integrating afforestation strategies with broader land-use policies to minimize negative impacts on food and land sustainability. However, it's essential to approach afforestation initiatives cautiously to avoid adverse consequences such as escalating land and food prices, which could exacerbate food insecurity (Mohan et al. 2021). Therefore, implementing afforestation alongside productive agricultural systems and safety-net programmes becomes imperative to mitigate associated risks effectively. While afforestation presents promising opportunities for carbon sequestration and ecosystem restoration, careful planning and integrated approaches are essential to maximize its benefits while minimizing negative impacts on food security and land sustainability. Afforestation and restoration efforts improve biological aerosol removal processes and increase the surface area available for aerosol deposition, which results in healthier surroundings and cleaner air.

5.3 Adaptation Strategies

Urban planning and design, public health initiatives, and agricultural practices are combinations that minimize exposure and vulnerability to aerosols as part of adaptation methods for aerosol pollution (Fig. 5.2). By using these tactics, vulnerable people will be shielded from the damaging impacts of aerosol pollution and their resilience will be strengthened.

5.3.1 Urban Planning and Design to Minimize Aerosol Exposure

Urban areas are particularly susceptible to aerosol pollution due to concentrated sources of emissions and limited dispersion. Urban planning and design initiatives can mitigate aerosol exposure by integrating green infrastructure, improving ventilation systems, and promoting sustainable development practices. Urban planning and design can play a crucial role in minimizing aerosol exposure (Li et al. 2018). The spatial variation of organic carbon (OC) and elemental carbon (EC) across different

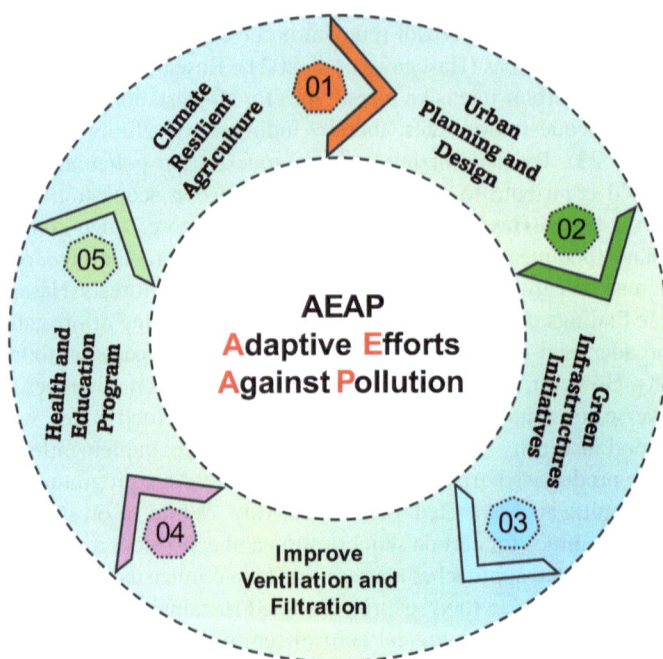

Fig. 5.2 Adaption efforts for extreme aerosol pollution

urban environments suggests that proximity to point sources significantly influences aerosol concentrations (Murillo et al. 2013). Particularly, OC concentrations increase near industrial areas and downtown locations, indicating the importance of zoning regulations and land-use planning to mitigate exposure to hazardous aerosols (US.EPA 1973; Li et al. 2018). Additionally, the association between different components of OC with thermal-optical analysis (OC_2 and OC_3) concentrations with traffic emissions, suggests the need for urban planners to prioritize measures that reduce vehicular traffic in densely populated areas (Cao et al. 2004; Gu et al. 2010). Strategies such as implementing green spaces, promoting public transportation, and designing pedestrian-friendly infrastructure can help reduce traffic congestion and subsequently lower aerosol emissions (Zarie et al. 2024). Furthermore, there is a need to consider both primary and secondary aerosol sources in urban planning efforts (Kanellopoulos et al. 2021). While primary emissions contribute to localized aerosol hotspots, secondary aerosols are more spatially homogeneous, emphasizing the need for comprehensive pollution control measures beyond point sources. Overall, urban planning and design interventions focused on reducing emissions from point sources, minimizing traffic congestion, and considering both primary and secondary aerosol sources can effectively mitigate aerosol exposure and improve public health in urban environments.

5.3.1.1 Green Infrastructure Initiatives (GIIs)

Green infrastructure (GI) initiatives offer promising avenues for adapting to and mitigating the impacts of air pollution in urban environments (Jayasooriya et al. 2017). The integration of green spaces, such as trees and parks, into urban landscapes has been shown to significantly reduce ambient temperatures and contribute to the removal of air pollutants (Diener and Mudu 2021). The analysis of various studies across different cities worldwide reveals that increased vegetation cover, particularly through the strategic planting of trees and the creation of green spaces, can lead to substantial decreases in ambient temperatures during both daytime and nighttime periods (Qiu et al. 2017; Yu et al. 2020; Santamouris and Osmond 2020). These temperature reductions are attributed to the cooling effects of evapotranspiration and shading provided by vegetation, as well as the decreased heat storage and sensible heat emissions from impervious surfaces like asphalt and concrete. Furthermore, the presence of green infrastructure can enhance air quality by trapping and removing pollutants from the atmosphere through processes such as dry deposition and filtration by vegetation (Santamouris and Osmond 2020). However, the effectiveness of GI initiatives in mitigating air pollution depends on factors such as climatic conditions, soil moisture availability, the type and distribution of vegetation, and the specific urban context (Wróblewska and Jeong 2021). While studies demonstrate the potential of GI to mitigate air pollution and improve urban microclimates, challenges such as urban planning considerations, uncertainty in modelling assumptions, and the need to address potential trade-offs with increased humidity levels highlight the complexity of implementing these initiatives (Santamouris and Osmond 2020; Sanusi and Jalil 2021; Ernst et al. 2022). Therefore, while GI offers valuable strategies for adaptation to air pollution, careful planning and consideration of local conditions are essential to maximize its effectiveness and minimize unintended consequences.

Additionally, the introduction and expansion of urban greenery, including trees and vegetation, have been identified as effective measures for reducing pollutant concentrations such as particulate matter (PM) and ground-level ozone (Diener and Mudu 2021). These initiatives leverage the dynamic interactions between vegetation and atmospheric pollutants, encompassing processes like particle deposition, pollutant absorption, and chemical reactions. Studies from diverse geographic locations such as New York, Melbourne, and Glasgow demonstrate the potential of increased urban green infrastructure in lowering pollutant concentrations. For example, in the Bronx, New York, an anticipated increase in tree cover by 2030 could lead to a substantial reduction in $PM_{2.5}$ levels, showcasing the effectiveness of strategic planning programmes (Nyelele et al. 2019). Similarly, simulations in Melbourne illustrate a remarkable increase in PM removal with additional trees, highlighting the role of green infrastructure in enhancing air quality (Santamouris and Osmond 2020). However, the impact of greenery on air pollution is nuanced, influenced by factors such as tree species, canopy structure, and meteorological conditions. While increased greenery can enhance particle removal through deposition processes, it may also lead to localized increases in pollutant concentrations due to reduced ventilation and atmospheric dispersion. Moreover, the complex chemistry between pollutants and vegetation emissions underscores the need for comprehensive modelling and assessment to inform effective green infrastructure planning. Therefore, green infrastructure initiatives must be carefully designed and implemented, considering local environmental conditions and pollutant dynamics, to maximize their benefits in adapting to and mitigating urban air pollution challenges.

Additionally, GIIs are essential for reducing the negative health effects of low air quality when it comes to air pollution adaptation techniques (Kumar et al. 2019). As explained in the above section, increasing urban greenery—such as trees and vegetation—helps to lower outside temperatures and lessen the effects of heat-related illness (Zhou et al. 2023). Studies from various cities, including Phoenix, Oslo, and Darwin, have demonstrated that the presence of green infrastructure can lead to a decrease in heat-related hospital admissions and emergency calls, particularly during periods of elevated temperatures (Silva et al. 2010; Venter et al. 2020; Yenneti et al. 2020). By enhancing natural cooling mechanisms and providing shade, green infrastructure helps alleviate the urban heat island effect, which is exacerbated by climate change and urbanization. Initiatives focusing on increasing green infrastructure not only offer immediate benefits in terms of air quality improvement and heat-related health risk reduction, but also contribute to building more resilient and sustainable urban environments in the face of climate change (Pamukcu-Albers et al. 2021). Therefore, investing in and expanding green infrastructure initiatives represents a crucial adaptation strategy for addressing air pollution-related health challenges in cities. Moreover, green roofs, installed on buildings, provide additional green space and help mitigate aerosol pollution by capturing particulate matter and reducing urban heat island effects. Permeable pavements allow rainwater to infiltrate the ground, reducing runoff and enhancing soil filtration of aerosols and pollutants.

5.3.1.2 Improved Ventilation and Filtration Systems in Buildings

Improved ventilation and filtration systems in buildings play a crucial role in mitigating the impact of aerosol pollution on indoor air quality and human health (Vijayan et al. 2015; Fan et al. 2022). The list of global air quality standards for indoor air quality is shown in Table 5.1. Traditional uniform and steady indoor environments, such as mixing ventilation (MV) and displacement ventilation (DV), have been widely employed to create a relatively stable indoor environment by diluting indoor pollutants (see Table 5.1) and providing efficient heat and mass exchange (Fan et al. 2022). However, these systems may not always guarantee optimal air quality, as they can lead to air short-circuiting, dead corners, and high energy consumption. Non-uniform ventilation modes, including personalized ventilation (PEV), underfloor air distribution (UFAD), and stratum ventilation (SV), offer more targeted approaches by delivering fresh air directly to the occupied zone, thereby improving thermal comfort and reducing energy consumption (Zhang et al. 2014). Additionally, advancements in ventilation technology, such as impinging jet ventilation (IJV) and wall-attached ventilation (WAV), provide efficient air distribution while minimizing draft sensation and enhancing air quality through reduced contaminant concentration (Karimipanah 2023). These systems offer promising solutions for maintaining indoor air quality in various settings, from office buildings to specialized environments like cabins and ice rinks, where specific air distribution requirements are essential for occupant comfort and safety.

Furthermore, the development of unsteady indoor environment systems, such as dynamic personalized ventilation (DPEV), pulsating ventilation (PUV), and inter- mittent ventilation (IV), offers dynamic control strategies to optimize indoor air quality while reducing energy consumption (Fan et al. 2022). These systems adapt to changing environmental conditions and occupant preferences by delivering clean air at different intervals and flow rates. Additionally, Wearable personal exhaust ventila- tion (WPEV) provides personalized protection against aerosol pollution, particularly in dynamic environments or scenarios where traditional ventilation systems may be impractical (Bolashikov et al. 2015). By integrating improved ventilation and filtra- tion technologies into building design and operation, such as enhanced air circulation, filtration efficiency, and air distribution control, it becomes possible to effectively mitigate the impact of aerosol pollution on indoor air quality, thereby promoting occupant health and well-being in various indoor environments.

5.3.2 Health and Education Programs for Vulnerable Populations

Vulnerable populations, including children, the elderly, and individuals with pre- existing health conditions, are disproportionately affected by aerosol pollution and require targeted health and education programmes to mitigate their exposure and health risks (Manisalidis et al. 2020). Efforts to raise awareness about the health risks associated with aerosol pollution have become increasingly important in urban areas (Kecorius et al. 2017). Through various awareness campaigns, policymakers and environmental organizations aim to educate the public about the detrimental effects of aerosol pollution on human health and the environment (Ramírez et al. 2019). These campaigns often utilize data from scientific studies to highlight the correlation between exposure to pollutants like particulate matter and respiratory diseases, cardiovascular problems, and other health issues. These initiatives aim to encourage people to take action to lessen their exposure and support policies that prioritize improving air quality by educating people on the causes and effects of aerosol pollution (Ramírez et al. 2019). Additionally, political efforts to decrease air pollution in urban areas involve the implementation of policies and initiatives aimed at reducing emissions from transportation, industry, and other sources (Mookherjee 2022; Shao and Chou 2023). These efforts may include the establishment of low- emission zones, investment in public transportation infrastructure, and the promotion of alternative modes of transportation such as cycling and walking. Furthermore,

Table 5.1 Global standards of the indoor air pollution

Pollutant	Global Standards for Indoor Air	Negative Effects
Carbon Dioxide (CO_2)	1000 ppm (Japanese Building Standards Act)	CO_2 levels can vary based on occupancy and ventilation. Elevated levels can cause drowsiness and decreased cognitive function
	1000 ppm (The American Society of Heating, Refrigerating and Air Conditioning Engineers)	
Carbon Monoxide (CO)	10 ppm (Japanese Building Standards Act & Building Hygiene Management Act)	CO is a colourless, odourless gas produced by incomplete combustion of fossil fuels. Exposure can lead to headaches, dizziness, and even death in high concentrations
	20 ppm (Japanese School Hygiene Standards)	
	9 ppm (US NAAQS: 8 Hour Average)	
	8.6 ppm (WHO Europe: 8 Hour Average)	
Formaldehyde (CH_2O)	0.1 ppm (The American Society of Heating, Refrigerating and Air Conditioning Engineers)	Formaldehyde is a common indoor air pollutant emitted by furniture, carpets, and household products. Prolonged exposure may cause respiratory issues and cancer
	0.08 ppm (WHO Europe)	
	0.08 ppm (Dubai Municipality, 8 Hour Average)	
Nitrogen Dioxide (NO2)	0.21 ppm (WHO Europe: 1 Hour Average)	NO2 is a byproduct of combustion processes, primarily from vehicles and industrial activities. It can exacerbate respiratory conditions and contribute to the formation of smog
	0.075 ppm (WHO Europe: 24 Hour Average)	
	0.053 ppm (NAAQS: 24 Hour Average)	
Radon	4.0 pCi/L (EPA)	Radon is a radioactive gas naturally occurring in soil and rock. It can seep into buildings and accumulate to dangerous levels, increasing the risk of lung cancer
	2.0 pCi/L (The American Society of Heating, Refrigerating and Air Conditioning Engineers)	
Total Suspended Particles / Particulate Matter	0.15 mg/m3 (Japanese Building Hygiene Management Act/Building Standards Act)	TSP/PM includes a mixture of aerosol particles suspended in the air, such as dust, pollen, and smoke. Inhalation of TSP/PM can irritate the respiratory system and exacerbate asthma and allergies
	0.1–0.12 mg/m3 (WHO: 8 Hour Average)	
	150 μg/m3 (Dubai Municipality, 8 Hour Average)	

(continued)

Table 5.1 (continued)

Pollutant	Global Standards for Indoor Air	Negative Effects
Volatile Organic Components (VOCs)	0.2–0.6 mg/m3 (FISIAQ (Finnish Society of Indoor Air Quality and Climate))	VOCs are emitted by various household products and building materials. Prolonged exposure may cause headaches, nausea, and respiratory irritation
	300 μg/m3 (Dubai Municipality, 8 Hour Average)	

governments may enact regulations to limit emissions from vehicles and industrial facilities, as well as invest in technologies and strategies to monitor and mitigate air pollution levels (Jonidi Jafari et al. 2021). Overall, awareness campaigns and political efforts play complementary roles in addressing aerosol pollution in urban areas, working towards the shared goal of safeguarding public health and environmental quality.

Political initiatives to reduce urban air pollution combine awareness campaigns with a multifaceted strategy that includes policy creation, infrastructure spending, and public involvement. Frameworks for international collaboration in addressing cross-border air pollution challenges are provided by conventions and agreements, such as the Convention on Long-range Transboundary Air Pollution (US Department of State). Governments enact laws and act at the municipal and national levels to lower emissions and raise air quality standards (Mookherjee 2022). These efforts often involve collaboration between government agencies, industry stakeholders, and community organizations to develop and implement effective strategies. For example, initiatives such as low-emission zones and investments in public transportation infrastructure aim to reduce reliance on private vehicles and promote sustainable mobility options (Nieuwenhuijsen 2020). Political initiatives aim to build more sustainable and healthful urban environments for present and future generations by combining focused legislative interventions with awareness-raising initiatives.

5.3.3 Climate-Resilient Agriculture Practices to Mitigate Aerosol Impacts on Crop Yield

Climate-resilient agriculture practices serve as a critical strategy to mitigate the impacts of aerosols on crop yield, recognizing the intricate interplay between climate, water, soil, and agricultural productivity (Karri and Nalluri 2024). Climate change is making weather patterns more unpredictable, which presents increasing issues for the agriculture industry, especially when it comes to guaranteeing food security (Raza et al. 2019). Rising temperatures, altered precipitation patterns, and increased climatic unpredictability pose significant threats to crop growth and yield, with each component of the ecosystem influencing the others in a complex cycle (Tripathi et al.

2016; Mall et al. 2017). Aerosol emissions, a consequence of industrial activities and natural phenomena like wildfires, further exacerbate these challenges, affecting plant growth directly and indirectly through altered weather conditions (Fan et al. 2015; Wang et al. 2023). From temperature fluctuations to changes in water availability and pest infestations, aerosols play a role in shaping the agricultural landscape, necessitating adaptive measures to ensure resilience in crop production.

Climate-resilient farming methods prioritize adaptive techniques at different phases of crop development, from crop selection to harvest, to solve these issues (Debangshi 2021). Smart crop and variety selection, informed by weather forecasts and long-term research, become paramount to mitigate the impacts of changing climatic conditions on crop yield (Rashid et al. 2022). Furthermore, efficient cropping systems, such as intercropping and mixed cropping, not only enhance productivity, but also contribute to soil health and weed control, thereby reducing vulnerabilities to aerosol-induced stressors (Kumawat et al. 2022). Additionally, climate-based agro-advisories and innovative techniques like precision farming and water harvesting offer valuable tools for farmers to navigate the complexities of a changing climate (Roy and George 2020). Stakeholders can ensure sustainable food production, reduce aerosol impacts, and increase resilience in the face of changing environmental problems by incorporating these techniques into agricultural systems.

5.4   Resilience Strategies

Resilience strategies are essential for enhancing societal preparedness and response capabilities in the face of aerosol pollution events (Fig. 5.3). These strategies aim to minimize the impacts of extreme aerosol events, such as dust storms and wildfires, protect vulnerable populations, and promote adaptive governance structures and innovative solutions.

5.4.1   Early Warning Systems for Extreme Aerosol Events

Early warning systems play a critical role in alerting communities to the onset of extreme aerosol events, enabling timely evacuation, emergency preparedness, and public health interventions (Song et al. 2015). These systems rely on remote sensing and monitoring technologies, as well as the integration of aerosol data into climate models, to improve forecasting accuracy and provide actionable information to decision-makers.

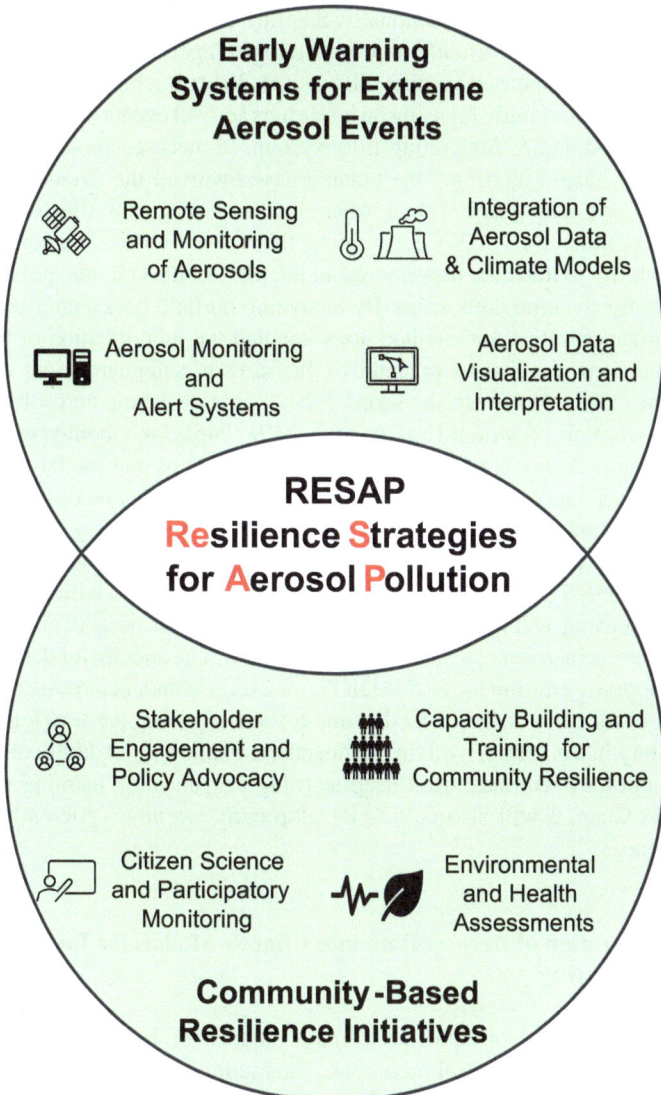

Fig. 5.3 Resilience strategies for extreme aerosol pollution

5.4.1.1 Remote Sensing and Monitoring Technologies

The development and application of early warning systems (EWS) for extreme aerosol events, particularly in the context of aviation safety, relies heavily on remote sensing and monitoring technologies (Papagiannopoulos et al. 2020). Remote sensing technologies, such as satellite imagery, LIDAR, and ground-based sensors, enable real-time monitoring of aerosol concentrations, air quality, and meteorological conditions (Ma et al. 2019). Additionally, Satellite-based sensors, such as MODIS, MERRA, and CALIPSO provide high-resolution imagery and vertical profiles of aerosol distributions in the atmosphere (Brun et al. 2011; Buchard et al. 2016; Chand et al. 2021). In a recent study focusing on hazardous aerosol events detected in Greece, advanced methodologies integrating lidar systems showcased promising capabilities for timely hazard alerting. One notable case involved the detection of desert dust particles originating from Libya, where intense dust storms affected the region (Papagiannopoulos et al. 2020). Leveraging lidar observations, the study demonstrated the ability to track the movement and intensity of dust clouds, providing critical insights for aviation authorities. By analyzing particle backscatter coefficients and depolarization ratios, the methodology enabled the identification of hazardous aerosol conditions, facilitating preemptive measures to safeguard flight operations (Zhang et al. 2021). However, the scope encompasses volcanic aerosols, exemplified by the eruption of Mount Etna in May 2019 (Papagiannopoulos et al. 2020). Through a comprehensive analysis of LIDAR observations and model simulations, the methodology successfully distinguished between volcanic and desert dust particles, crucial for accurate hazard assessment. This interdisciplinary approach, integrating real-time lidar data with atmospheric modelling, holds significant promise for enhancing EWS capabilities. Through the utilization of modelling frameworks and remote sensing technology, aviation stakeholders can proactively reduce the impact of extreme aerosol events on flight operations. The successful demonstration of this methodology during the EUNADICS-AV exercise underscores its potential as a reliable tool for early hazard detection and response, paving the way for enhanced aviation safety in the face of evolving atmospheric challenges. Additionally, remote sensing technology with advanced deep learning and machine learning models as discussed in Chap. 3 will also help to develop early warning systems for a short period of time.

5.4.1.2 Integration of Aerosol Data into Climate Models for Improved Forecasting

The integration of aerosol data into climate models for improved forecasting is paramount for understanding the complex interactions between aerosols and the Earth's radiation budget (Choi et al. 2019). The impact of aerosol climatology on atmospheric radiative fluxes and temperatures highlights the implementation of a new aerosol climatology in global forecasting models significantly influenced clear-sky shortwave downward radiative fluxes, leading to a reduction in surface temperatures globally (Bozzo et al. 2020). Notably, heavy aerosol-loading regions, such as the Sahara, Arabian Peninsula, and western India experienced substantial reductions in radiative flux and surface temperatures (Kedia et al. 2018; Ravi Kumar et al. 2019). This reduction in shortwave radiative flux due to aerosol direct effect was accompanied by enhanced shortwave heating rates in the troposphere, highlighting the intricate relationship between aerosols and atmospheric heating dynamics. Additionally, the consideration of aerosol climatology led to changes in cloud properties, including reduced low-level cloud fraction, which further influenced atmospheric heating and cooling patterns (Barthlott et al. 2022).

Furthermore, medium-range forecasts revealed the significant impact of aerosol data integration on forecast skill scores for various meteorological variables (Jeong 2020). The inclusion of aerosol direct effects in the forecasting model resulted in improved temperature and specific humidity forecasts, particularly in regions with heavy aerosol loading (Zhao et al. 2024). Additionally, statistical skill scores for precipitation forecasts showed enhancements, with a closer spatial pattern resemblance to observations, particularly over Asia and tropical regions (Choi et al. 2019). The reduced precipitation in the forecasting model was attributed to weakened convection intensity induced by aerosol-induced atmospheric heating and reduced relative humidity in the low-level troposphere. Overall, the integration incorporates data into climate models not only improves our understanding of aerosol-radiation interactions, but also enhances the accuracy of medium-range forecasts, offering valuable insights for climate prediction and adaptation strategies. Additionally, numerical weather prediction (NWP) models, such as the Weather Research and Forecasting (WRF) model, incorporate aerosol data to simulate the interactions between aerosols and meteorological variables, such as wind speed, humidity, and atmospheric stability (Qiu et al. 2023).

5.4.2 Community-Based Resilience Initiatives

Community-based initiatives addressing aerosol pollution have emerged as critical drivers of resilience and empowerment (Ward et al. 2022). Predominantly conducted in the USA, Europe, and Canada, these initiatives often centred around engaging local communities in activities related to outdoor air pollution (Ward et al. 2022). Ranging from participatory monitoring programmes to educational campaigns and policy development efforts, these initiatives aimed to raise awareness, empower individuals and communities, and generate local knowledge to support decision-making (Hilhorst and Guijt 2006). Notably, many studies highlighted the disproportionate exposure to pollution among disadvantaged and marginalized groups, framing aerosol pollution within the context of health inequalities (Pasa 2017; Phillip et al. 2023; Karri and Nalluri 2024). These programmes encouraged a sense of ownership, empowerment, and community resilience in the fight against aerosol pollution by actively integrating locals in all stages of the process, such as data collection, analysis, and policy advocacy. However, challenges such as language barriers, capacity limitations, and trust issues with authorities posed significant hurdles to effective community engagement, underscoring the importance of tailored approaches and equitable partnerships to ensure inclusivity and effectiveness in addressing aerosol pollution at the grassroots level.

One of the most prevalent approaches observed in these initiatives was citizen science, where residents actively participated in monitoring air quality using scientific tools and methodologies (Mahajan et al. 2020). This citizen-led monitoring not only provided valuable data, but also empowered communities to take proactive measures in addressing local air pollution issues. Additionally, community engagement extended beyond mere data collection to encompass environmental and health assessments, education and training programmes, and policy development efforts (Haldane et al. 2019). Through these multifaceted approaches, individuals and communities gained a deeper understanding of aerosol pollution and its health impacts, developed essential skills and competencies, and actively contributed to shaping policies and practices aimed at improving air quality. Moreover, community engagement often catalyzed changes within statutory organizations, leading to policy reforms, enhanced funding, and improved responsiveness to community concerns. While the direct impact on air quality and health outcomes was seldom measured, the collective efforts of these community-based resilience initiatives laid a foundation for sustainable action and empowered individuals to safeguard their well-being in the face of aerosol pollution challenges.

5.5 Conclusion

Addressing the complexities of aerosol pollution requires a multifaceted approach that integrates various approaches, including mitigation, adaptation, and resilience. Mitigation efforts focus on reducing anthropogenic emissions through policies targeting key sources and technological advancements in cleaner production processes. Carbon capture and storage technologies offer promising avenues for capturing and sequestering aerosols, while nature-based solutions enhance natural aerosol removal processes. Urban planning initiatives, including green infrastructure and improved ventilation systems, mitigate aerosol exposure in urban areas. Health and education programmes raise awareness and protect vulnerable populations from aerosol-related health risks, while climate-resilient agriculture practices enhance crop yield resilience.

Adaptation strategies emphasize early warning systems for extreme aerosol events, integrating remote sensing technologies and aerosol data into climate models to improve forecasting accuracy. Community-based resilience initiatives empower local communities to address aerosol pollution through citizen science and policy advocacy. However, challenges such as capacity limitations and trust issues underscore the importance of tailored approaches and equitable partnerships.

Additionally, policy interventions are crucial for effective aerosol pollution management. These may include establishing emission standards, incentivizing cleaner technologies, and investing in sustainable infrastructure. Additionally, fostering international collaboration and promoting community engagement can enhance resilience and empower individuals to address aerosol pollution at the grassroots level. By implementing a comprehensive array of strategies and interventions, we can mitigate the adverse impacts of aerosol pollution, protect public health, and build more resilient communities and ecosystems for the future.

References

Ajayi T, Gomes JS, Bera A (2019) A review of CO_2 storage in geological formations emphasizing modeling, monitoring and capacity estimation approaches. Pet Sci 16:1028–1063. https://doi.org/10.1007/s12182-019-0340-8

Bajpai S, Shreyash N, Singh S et al (2022) Opportunities, challenges and the way ahead for carbon capture, utilization and sequestration (CCUS) by the hydrocarbon industry: Towards a sustainable future. Energy Rep 8:15595–15616. https://doi.org/10.1016/j.egyr.2022.11.023

Bandilla KW (2020) Carbon capture and storage. In: Future Energy. Elsevier, pp 669–692

Barthlott C, Zarboo A, Matsunobu T, Keil C (2022) Importance of aerosols and shape of the cloud droplet size distribution for convective clouds and precipitation. Atmos Chem Phys 22:2153–2172. https://doi.org/10.5194/acp-22-2153-2022

Barwise Y, Kumar P (2020) Designing vegetation barriers for urban air pollution abatement: a practical review for appropriate plant species selection. npj Clim Atmos Sci 3:12. https://doi.org/10.1038/s41612-020-0115-3

Bolashikov ZD, Barova M, Melikov AK (2015) Wearable personal exhaust ventilation: improved indoor air quality and reduced exposure to air exhaled from a sick doctor. Sci Technol Built Environ 21:1117–1125. https://doi.org/10.1080/23744731.2015.1091270

Bozzo A, Benedetti A, Flemming J et al (2020) An aerosol climatology for global models based on the tropospheric aerosol scheme in the integrated forecasting system of ECMWF. Geosci Model Dev 13:1007–1034. https://doi.org/10.5194/gmd-13-1007-2020

Brun J, Shrestha P, Barros AP (2011) Mapping aerosol intrusion in Himalayan valleys using the Moderate Resolution Imaging Spectroradiometer (MODIS) and Cloud-Aerosol Lidar and Infrared Pathfinder Satellite Observation (CALIPSO). Atmos Environ 45:6382–6392. https://doi.org/10.1016/j.atmosenv.2011.08.026

Buchard V, da Silva AM, Randles CA et al (2016) Evaluation of the surface PM2.5 in Version 1 of the NASA MERRA aerosol reanalysis over the united states. Atmos Environ 125:100–111. https://doi.org/10.1016/j.atmosenv.2015.11.004

Cao J, Lee S, Ho K et al (2004) Spatial and seasonal variations of atmospheric organic carbon and elemental carbon in Pearl River Delta Region, China. Atmos Environ 38:4447–4456. https://doi.org/10.1016/j.atmosenv.2004.05.016

Chakraborty S, Guan B, Waliser DE, et al (2021) Extending the atmospheric river concept to aerosols: climate and air quality impacts. Geophys Res Lett 48.https://doi.org/10.1029/2020GL091827

Chand K, Kuniyal JC, Kanga S et al (2021) Aerosol characteristics and their impact on the Himalayan energy budget. Sustainability 14:179. https://doi.org/10.3390/su14010179

Choi I-J, Park R-S, Lee J (2019) Impacts of a newly-developed aerosol climatology on numerical weather prediction using a global atmospheric forecasting model. Atmos Environ 197:77–91. https://doi.org/10.1016/j.atmosenv.2018.10.019

Cunningham SC, Mac Nally R, Baker PJ et al (2015) Balancing the environmental benefits of reforestation in agricultural regions. Perspect Plant Ecol Evol Syst 17:301–317. https://doi.org/10.1016/j.ppees.2015.06.001

D'Alessandro DM, Smit B, Long JR (2010) Carbon dioxide capture: prospects for new materials. Angew Chemie Int Ed 49:6058–6082. https://doi.org/10.1002/anie.201000431

Debangshi U (2021) Climate resilient agriculture—An approach to reduce the ill-effect of climate change. Int J Recent Adv Multidiscip Top 2. https://doi.org/10.5281/zenodo.5545934

Diener A, Mudu P (2021) How can vegetation protect us from air pollution? A critical review on green spaces' mitigation abilities for air-borne particles from a public health perspective—with implications for urban planning. Sci Total Environ 796:148605. https://doi.org/10.1016/j.scitotenv.2021.148605

Dubey S, Goyal MK (2020) Glacial lake outburst flood hazard, downstream impact, and risk over the Indian Himalayas. Water Resour Res 56.https://doi.org/10.1029/2019WR026533

Ernst M, Le Mentec S, Louvrier M, et al (2022) Impact of urban greening on microclimate and air quality in the urban canopy layer: Identification of knowledge gaps and challenges. Front Environ Sci 10:https://doi.org/10.3389/fenvs.2022.924742

Fan J, Rosenfeld D, Yang Y et al (2015) Substantial contribution of anthropogenic air pollution to catastrophic floods in Southwest China. Geophys Res Lett 42:6066–6075. https://doi.org/10.1002/2015GL064479

Fan M, Fu Z, Wang J et al (2022) A review of different ventilation modes on thermal comfort, air quality and virus spread control. Build Environ 212:108831. https://doi.org/10.1016/j.buildenv.2022.108831

Fuller R, Landrigan PJ, Balakrishnan K et al (2022) Pollution and health: a progress update. Lancet Planet Heal 6:e535–e547. https://doi.org/10.1016/S2542-5196(22)00090-0

Gao L, Li H (2023) Improving carbon sequestration capacity of forest vegetation in china: afforestation or forest management? Forests 14:1077. https://doi.org/10.3390/f14061077

García-Bordejé E, González-Olmos R (2024) Advances in process intensification of direct air CO_2 capture with chemical conversion. Prog Energy Combust Sci 100:101132. https://doi.org/10.1016/j.pecs.2023.101132

Gu J, Bai Z, Liu A et al (2010) Characterization of atmospheric organic carbon and element carbon of PM2.5 and PM10 at Tianjin. China. Aerosol Air Qual Res 10:167–176. https://doi.org/10.4209/aaqr.2009.12.0080

Haldane V, Chuah FLH, Srivastava A et al (2019) Community participation in health services development, implementation, and evaluation: a systematic review of empowerment, health, community, and process outcomes. PLoS ONE 14:e0216112. https://doi.org/10.1371/journal.pone.0216112

Hasegawa T, Fujimori S, Ito A, Takahashi K (2024) Careful selection of forest types in afforestation can increase carbon sequestration by 25% without compromising sustainability. Commun Earth Environ 5:171. https://doi.org/10.1038/s43247-024-01336-4

Hilhorst T, Guijt I (2006) Participatory monitoring and evaluation : a process to support governance and empowerment at the local level. Using Particip Monit Eval to strengthen local Gov 54

Hu J, Guo Z, Chu W et al (2015) Carbon dioxide catalytic conversion to nano carbon material on the iron–nickel catalysts using CVD-IP method. J Energy Chem 24:620–625. https://doi.org/10.1016/j.jechem.2015.09.006

Jayasooriya VM, Ng AW, Muthukumaran S, Perera BJC (2017) Green infrastructure practices for improvement of urban air quality. Urban for Urban Green 21:34–47. https://doi.org/10.1016/j.ufug.2016.11.007

Jeong G-R (2020) Weather effects of aerosols in the global forecast model. Atmosphere (basel) 11:850. https://doi.org/10.3390/atmos11080850

Jiang XQ, Mei XD, Feng D (2016) Air pollution and chronic airway diseases: what should people know and do? J Thorac Dis 8:E31–E40. https://doi.org/10.3978/J.ISSN.2072-1439.2015.11.50

Jonidi Jafari A, Charkhloo E, Pasalari H (2021) Urban air pollution control policies and strategies: a systematic review. J Environ Heal Sci Eng 19:1911–1940. https://doi.org/10.1007/s40201-021-00744-4

Kanellopoulos PG, Verouti E, Chrysochou E et al (2021) Primary and secondary organic aerosol in an urban/industrial site: Sources, health implications and the role of plastic enriched waste burning. J Environ Sci 99:222–238. https://doi.org/10.1016/j.jes.2020.06.012

Karimipanah T (2023) Some Aspects of HVAC design in energy renovation of buildings. In: Urban transition—perspectives on urban systems and environments. IntechOpen

Karri V, Nalluri N (2024) Enhancing resilience to climate change through prospective strategies for climate-resilient agriculture to improve crop yield and food security. Plant Sci Today 11:21–33. https://doi.org/10.14719/pst.2140

Kätelhön A, Meys R, Deutz S et al (2019) Climate change mitigation potential of carbon capture and utilization in the chemical industry. Proc Natl Acad Sci 116:11187–11194. https://doi.org/10.1073/pnas.1821029116

Kaur R, Pandey P (2021) Air pollution, climate change, and human health in indian cities: a brief review. Front Sustain Cities 3https://doi.org/10.3389/frsc.2021.705131

Kecorius S, Madueño L, Vallar E et al (2017) Aerosol particle mixing state, refractory particle number size distributions and emission factors in a polluted urban environment: Case study of Metro Manila, Philippines. Atmos Environ 170:169–183. https://doi.org/10.1016/j.atmosenv.2017.09.037

Kedia S, Kumar R, Islam S et al (2018) Radiative impact of a heavy dust storm over India and surrounding oceanic regions. Atmos Environ 185:109–120. https://doi.org/10.1016/j.atmosenv.2018.05.005

Kelly FJ, Fussell JC (2015) Air pollution and public health: emerging hazards and improved understanding of risk. Environ Geochem Health 37:631–649. https://doi.org/10.1007/s10653-015-9720-1

Ketzer JM, Iglesias RS, Einloft S (2012) Reducing greenhouse gas emissions with CO_2 capture and geological storage. Handbook of Climate Change Mitigation. Springer, US, New York, NY, pp 1405–1440

Kim G, Kim J, Ko Y et al (2021) How do nature-based solutions improve environmental and socio-economic resilience to achieve the sustainable development goals? Reforestation and Afforestation Cases from the Republic of Korea. Sustainability 13:12171. https://doi.org/10.3390/su132112171

Kumar N, Poonia V, Gupta BB, Goyal MK (2021) A novel framework for risk assessment and resilience of critical infrastructure towards climate change. Technol Forecast Soc Change 165:120532. https://doi.org/10.1016/j.techfore.2020.120532

Kumar P, Druckman A, Gallagher J et al (2019) The nexus between air pollution, green infrastructure and human health. Environ Int 133:105181. https://doi.org/10.1016/j.envint.2019.105181

Kumawat A, Bamboriya SD, Meena RS et al (2022) Legume-based inter-cropping to achieve the crop, soil, and environmental health security. In: Advances in legumes for sustainable intensification. Elsevier, pp 307–328

Li HZ, Dallmann TR, Li X et al (2018) Urban organic aerosol exposure: spatial variations in composition and source impacts. Environ Sci Technol 52:415–426. https://doi.org/10.1021/acs.est.7b03674

Li N, Mo L, Unluer C (2022) Emerging CO_2 utilization technologies for construction materials: a review. J CO_2 Util 65:102237. https://doi.org/10.1016/j.jcou.2022.102237

Liu D, He C, Schwarz JP, Wang X (2020) Lifecycle of light-absorbing carbonaceous aerosols in the atmosphere. npj Clim Atmos Sci 3:40. https://doi.org/10.1038/s41612-020-00145-8

Lu X, Zhang S, Xing J et al (2020) Progress of air pollution control in china and its challenges and opportunities in the ecological civilization era. Engineering 6:1423–1431. https://doi.org/10.1016/j.eng.2020.03.014

Ma X, Wang C, Han G et al (2019) Regional atmospheric aerosol pollution detection based on LiDAR remote sensing. Remote Sens 11:2339. https://doi.org/10.3390/rs11202339

Mahajan S, Kumar P, Pinto JA et al (2020) A citizen science approach for enhancing public understanding of air pollution. Sustain Cities Soc 52:101800. https://doi.org/10.1016/j.scs.2019.101800

Mall RK, Gupta A, Sonkar G (2017) Effect of climate change on agricultural crops. In: Current developments in biotechnology and bioengineering. Elsevier, pp 23–46

Manisalidis I, Stavropoulou E, Stavropoulos A, Bezirtzoglou E (2020) Environmental and health impacts of air pollution: a review. Front Public Heal 8.https://doi.org/10.3389/fpubh.2020.00014

Mo, Zhang, Li, Qu (2019) A novel air quality early-warning system based on artificial intelligence. Int J Environ Res Public Health 16:3505.https://doi.org/10.3390/ijerph16193505

Mohan M, Rue HA, Bajaj S et al (2021) Afforestation, reforestation and new challenges from COVID-19: Thirty-three recommendations to support civil society organizations (CSOs). J Environ Manage 287:112277. https://doi.org/10.1016/j.jenvman.2021.112277

Mookherjee P (2022) India's air pollution challenge: translating policies into effective action

Murillo JH, Rojas Marin JF, Roman SR et al (2013) Temporal and spatial variations in organic and elemental carbon concentrations in PM10/PM2.5 in the metropolitan area of Costa Rica. Central America. Atmos Pollut Res 4:53–63. https://doi.org/10.5094/APR.2013.006

Nagireddi S, Agarwal JR, Vedapuri D (2024) Carbon dioxide capture, utilization, and sequestration: current status, challenges, and future prospects for global decarbonization. ACS Eng Au 4:22–48. https://doi.org/10.1021/acsengineeringau.3c00049

Nieuwenhuijsen MJ (2020) Urban and transport planning pathways to carbon neutral, liveable and healthy cities; a review of the current evidence. Environ Int 140:105661. https://doi.org/10.1016/j.envint.2020.105661

Nyelele C, Kroll CN, Nowak DJ (2019) Present and future ecosystem services of trees in the Bronx, NY. Urban for Urban Green 42:10–20. https://doi.org/10.1016/j.ufug.2019.04.018

Pamukcu-Albers P, Ugolini F, La Rosa D et al (2021) Building green infrastructure to enhance urban resilience to climate change and pandemics. Landsc Ecol 36:665–673. https://doi.org/10.1007/s10980-021-01212-y

Papagiannopoulos N, D'Amico G, Gialitaki A et al (2020) An EARLINET early warning system for atmospheric aerosol aviation hazards. Atmos Chem Phys 20:10775–10789. https://doi.org/10.5194/acp-20-10775-2020

Park S, Allen RJ, Lim CH (2020) A likely increase in fine particulate matter and premature mortality under future climate change. Air Qual Atmos Heal 13:143–151. https://doi.org/10.1007/s11869-019-00785-7

Pasa RB (2017) Role of capacity/skill development trainings in rural livelihood: a case study of Hapur, Dang. J Train Dev 3:41–49. https://doi.org/10.3126/jtd.v3i0.18229

Phillip E, Langevin J, Davis M et al (2023) Improved cookstoves to reduce household air pollution exposure in sub-Saharan Africa: a scoping review of intervention studies. PLoS ONE 18:e0284908. https://doi.org/10.1371/journal.pone.0284908

Priyadarshini P, Rim G, Rosu C et al (2023) Direct Air capture of CO_2 using Amine/Alumina Sorbents at cold temperature. ACS Environ Au 3:295–307. https://doi.org/10.1021/acsenvironau.3c00010

Qiu GY, Zou Z, Li X et al (2017) Experimental studies on the effects of green space and evapotranspiration on urban heat island in a subtropical megacity in China. Habitat Int 68:30–42. https://doi.org/10.1016/j.habitatint.2017.07.009

Qiu Y, Feng J, Zhang Z, et al (2023) Regional aerosol forecasts based on deep learning and numerical weather prediction. npj Clim Atmos Sci 6:71. https://doi.org/10.1038/s41612-023-00397-0

Ramírez AS, Ramondt S, Van Bogart K, Perez-Zuniga R (2019) Public awareness of air pollution and health threats: challenges and opportunities for communication strategies to improve environmental health literacy. J Health Commun 24:75–83. https://doi.org/10.1080/10810730.2019.1574320

Rashid S, Bin Mushtaq M, Farooq I, Khan Z (2022) Climate smart crops for food security. In: The nature, causes, effects and mitigation of climate change on the environment. IntechOpen

Rautela KS, Singh S, Goyal MK (2024a) Characterizing the spatio-temporal distribution, detection, and prediction of aerosol atmospheric rivers on a global scale. J Environ Manage 351:119675. https://doi.org/10.1016/j.jenvman.2023.119675

Rautela KS, Singh S, Goyal MK (2024b) Resilience to air pollution: A novel approach for detecting and predicting aerosol atmospheric rivers within earth system boundaries earth systems and environment. https://doi.org/10.1007/s41748-024-00421-0

Ravi Kumar K, Attada R, Dasari HP et al (2019) On the recent amplification of dust over the Arabian peninsula during 2002–2012. J Geophys Res Atmos 124:13220–13229. https://doi.org/10.1029/2019JD030695

Raza A, Razzaq A, Mehmood S et al (2019) Impact of climate change on crops adaptation and strategies to tackle its outcome: a review. Plants 8:34. https://doi.org/10.3390/plants8020034

Roy T, George K J (2020) Precision farming: a step towards sustainable, climate-smart agriculture. In: Global climate change: resilient and smart agriculture. Springer Singapore, Singapore, pp 199–220

Santamouris M, Osmond P (2020) Increasing green infrastructure in cities: impact on ambient temperature, air quality and heat-related mortality and morbidity. Buildings 10:233. https://doi.org/10.3390/buildings10120233

Sanusi R, Jalil M (2021) Blue-Green infrastructure determines the microclimate mitigation potential targeted for urban cooling. IOP Conf Ser Earth Environ Sci 918:012010. https://doi.org/10.1088/1755-1315/918/1/012010

Sanz-Pérez ES, Murdock CR, Didas SA, Jones CW (2016) Direct capture of CO_2 from ambient air. Chem Rev 116:11840–11876. https://doi.org/10.1021/acs.chemrev.6b00173

Shammin MR, Haque AKE, Faisal IM (2022) A framework for climate resilient community-based adaptation. In: Climate change and community resilience. Springer Nature Singapore, Singapore, pp 11–30

Shao W-C, Chou L-C (2023) Political influence and air pollution: evidence from Chinese cities. Heliyon 9:e17781. https://doi.org/10.1016/j.heliyon.2023.e17781

Shaw R, Mukherjee S (2022) The development of carbon capture and storage (CCS) in India: a critical review. Carbon Capture Sci Technol 2:100036. https://doi.org/10.1016/j.ccst.2022.100036

Silva HR, Phelan PE, Golden JS (2010) Modeling effects of urban heat island mitigation strategies on heat-related morbidity: a case study for Phoenix, Arizona, USA. Int J Biometeorol 54:13–22. https://doi.org/10.1007/s00484-009-0247-y

Sneddon G, Greenaway A, Yiu HHP (2014) The potential applications of nanoporous materials for the adsorption, separation, and catalytic conversion of carbon dioxide. Adv Energy Mater 4.https://doi.org/10.1002/aenm.201301873

Song Y, Qin S, Qu J, Liu F (2015) The forecasting research of early warning systems for atmospheric pollutants: a case in Yangtze River Delta region. Atmos Environ 118:58–69. https://doi.org/10.1016/j.atmosenv.2015.06.032

Southerland VA, Brauer M, Mohegh A et al (2022) Global urban temporal trends in fine particulate matter (PM2·5) and attributable health burdens: estimates from global datasets. Lancet Planet Heal 6:e139–e146. https://doi.org/10.1016/S2542-5196(21)00350-8

State USD of Convention on Long-Range Transboundary Air Pollution—United States Department of State. https://www.state.gov/key-topics-office-of-environmental-quality-and-transboundary-issues/convention-on-long-range-transboundary-air-pollution/. Accessed 15 Apr 2024

Sturiale S (2019) The role of green infrastructures in urban planning for climate change adaptation. Climate 7:119. https://doi.org/10.3390/cli7100119

Tripathi A, Tripathi DK, Chauhan DK et al (2016) Paradigms of climate change impacts on some major food sources of the world: a review on current knowledge and future prospects. Agric Ecosyst Environ 216:356–373. https://doi.org/10.1016/j.agee.2015.09.034

US.EPA (1973) A guide for reducing airpollution through urban planning.pdf

Venter ZS, Krog NH, Barton DN (2020) Linking green infrastructure to urban heat and human health risk mitigation in Oslo. Norway. Sci Total Environ 709:136193. https://doi.org/10.1016/j.scitotenv.2019.136193

Vijayan V, Paramesh H, Salvi S, Dalal AK (2015) Enhancing indoor air quality—The air filter advantage. Lung India 32:473. https://doi.org/10.4103/0970-2113.164174

Wang P, Yang Y, Xue D et al (2023) Aerosols overtake greenhouse gases causing a warmer climate and more weather extremes toward carbon neutrality. Nat Commun 14:7257. https://doi.org/10.1038/s41467-023-42891-2

Wang X, Zhang F, Li L, et al (2021) Carbon dioxide capture

Ward F, Lowther-Payne HJ, Halliday EC et al (2022) Engaging communities in addressing air quality: a scoping review. Environ Heal 21:89. https://doi.org/10.1186/s12940-022-00896-2

Wróblewska K, Jeong BR (2021) Effectiveness of plants and green infrastructure utilization in ambient particulate matter removal. Environ Sci Eur 33:110. https://doi.org/10.1186/s12302-021-00547-2

Yenneti K, Ding L, Prasad D et al (2020) Urban Overheating and cooling potential in Australia: an evidence-based review. Climate 8:126. https://doi.org/10.3390/cli8110126

Yu Z, Yang G, Zuo S et al (2020) Critical review on the cooling effect of urban blue-green space: a threshold-size perspective. Urban for Urban Green 49:126630. https://doi.org/10.1016/j.ufug.2020.126630

Zarie E, Sepehri B, Adibhesami MA et al (2024) A strategy for giving urban public green spaces a third dimension: a case study of Qasrodasht. Shiraz. Nature-Based Solut 5:100102. https://doi.org/10.1016/j.nbsj.2023.100102

Zhang K, Zhang X, Li S, Jin X (2014) Review of underfloor air distribution technology. Energy Build 85:180–186. https://doi.org/10.1016/j.enbuild.2014.09.011

Zhang X, Wang H, Che H-Z et al (2021) Radiative forcing of the aerosol-cloud interaction in seriously polluted East China and East China Sea. Atmos Res 252:105405. https://doi.org/10.1016/j.atmosres.2020.105405

Zhao M, Dai T, Goto D et al (2024) Assessing the assimilation of Himawari-8 observations on aerosol forecasts and radiative effects during pollution transport from South Asia to the Tibetan Plateau. Atmos Chem Phys 24:235–258. https://doi.org/10.5194/acp-24-235-2024

Zhao T, Guo B, Li Q et al (2016) Highly efficient CO_2 capture to a new-style CO_2-storage material. Energy Fuels 30:6555–6560. https://doi.org/10.1021/acs.energyfuels.6b00443

Zhou W, Yu W, Zhang Z et al (2023) How can urban green spaces be planned to mitigate urban heat island effect under different climatic backgrounds? A threshold-based perspective. Sci Total Environ 890:164422. https://doi.org/10.1016/j.scitotenv.2023.164422